Anthropometrics for Designers

First published 1971
7134 0524 4
Library of Congress Catalog Card Number 73-130690

Filmset in Great Britain by Filmtype Services Limited Scarborough Yorkshire. Printed in Great Britain by offset litho at Taylor Garnett Evans & Co Ltd Watford Herts and bound by Wm Brendon & Son Ltd Tiptree Essex for the publishers B T Batsford Limited 4 Fitzhardinge Street London W1 and Van Nostrand Reinhold Company a Division of Litton Educational Publishing Inc. 450 West 33rd Street New York NY 10001

Anthropometrics for Designers

John Croney

B. T. Batsford Ltd *London*

Van Nostrand Reinhold Company *New York*

Contents

Dedicated to Isobel, Mary and Sarah.

Acknowledgment

I wish to acknowledge the help I have had from many people and sources in the course of the preparation of this book.

Published anatomical and anthropological information has been the source of some of the facts used, other information has come from experts who have shared their knowledge and experience.

I am grateful to Professor J. N. Tanner and to his publishers Blackwell Scientific Publications, Oxford, for allowing me to use material from Growth at Adolescence as a basis for my own findings for this book. Illustration **3** was derived in part from a photograph showing the proportions of boys' faces in the above work. The idea of the fitted curve showing incremental increases in growth shown in Illustrations **14a** and **14b** has been derived from the same source. The measurements have been calculated by me and for the purpose of clarity stated as simple numbers.

I am grateful to Dr. W. H. Sheldon and his publishers Harper and Row, New York, for allowing me to use photographs of male somatotypes from the Atlas of Men to construct the examples shown in Illustrations **26**, **27** and **28** and for the use of the information in The Varieties of Human Physique on the incidence of somatotypes in a population. This information forms the basis for Chart **10**.

My thanks are due to my colleagues who have freely made available their ideas, and who have directed my attention towards certain aspects of human measurement I may otherwise have missed. They are due as well to Mr. Robert Arnold for the generous and lucid manner with which he shared biological knowledge. Also to Mr. A. E. Halliwell and the teaching staff of the Industrial Design Department at the Central School of Art and Design for allowing me to benefit from their experience gained when designing equipment for a variety of human tasks. I am grateful for advice and specifications given by Mr. Nigel Walters on the design of seats and domestic furniture in relation to human measurements and use.

Any errors in the use of the measurements and information I have been allowed to use are entirely my own, as are the conclusions I have reached from them.

In the field of practical research I must thank all the models who patiently assisted my study of human measurement and physique.

I must express my debt to Miss B. J. Kirkpatrick and the Library staff of the Royal Anthropological Institute; to Mrs. K. Ely of the Croydon Library service and to Miss A. M. Coghlan, librarian of the College of Fashion and Clothing Technology who have gone to much trouble to obtain material of use.

The preparation of the manuscript was made much easier by the typing and assistance of Miss E. Ridley in the initial stage; and later by Miss M. Leach.

My greatest debt is to those at Batsfords who were concerned with the publication of the book and the personal interest they have shown in it: to Mr. Guy Warren for the design of the layout; to Miss Mary Scriven who made liaison arrangements so easy and pleasant; to Miss Thelma Nye for her help with metrication, and lastly to Mr. Samuel Carr for his invaluable guidance, unfailing enthusiasm and kindness.

Introduction

I have no faith in anything short of actual Measurement
Charles Darwin

Nothing seems quite so permanent as man in the contemporary situation, despite all we say to the contrary. He seems to be the one constant factor amongst the ever increasing amount of inventions with which he hedges himself in. It is impossible to correlate artistic and industrial efforts to help manufacture things of use, without gathering information about man first of all. If the information is to be worthwhile and apposite one of the things man must do is to measure himself. It is this necessary concern of man with himself, and the tools he wants to use, that is the justification for this book.

To know about man is to know a great deal about the possible environments in which he can function satisfactorily; and in fact the most successful environmental system would be a replica of his own system. That is a system that could function in the same atmosphere, at the same temperature, etc. To be a successful designer of products for a part, or the whole, of a human environment, a designer must be conversant with different types of human physique and be informed about the limitations of human performance. Also now, more than ever before, man demands products that are tailor-made to suit specific requirements.

This book has been designed to provide an illustrated account of man's dimensions and other physical data, and to define their limitations and comment on any peculiarities. It is intended as a reference for use by designers in specialised fields of industrial or commercial design (for example in the dress and furniture trades), and by engineers, architects and builders.

If we can consider that the work of many designers is concerned with solving problems of environment it is hoped that the book will help towards the invention of fresh solutions. The contents deal with the physical attributes and shapes of man; bio-chemical processes; and changes in the body's physiological systems which may be attributable to ageing or environment are not dealt with, and any attempt to have done so would have been contrary to the purpose of the book.

One of the more important aspects of this type of book is the consideration of the types of language that should be used to describe the facts and ideas it contains. A great deal of anthropometric information can be conveyed visually; and a scheme consisting of a series of diagrams was decided upon.

The text has been kept as clear and simple as possible, and the aim has been to achieve a pertinent association of words and diagrams. In this way a course has been steered between a text overloaded with natural-science language, or one that is either too extravagantly personal or too scanty and journalistic.

When the aim is to be useful and comprehensive one cannot be original as well. Many of the facts used are not my own; but they are also not generally available. A great many have been collected from the technical literature of Europe and America, and I have also taken facts from the books listed in the Bibliography. I have incorporated much current thinking in anthropometry, human biology and anatomy and have made the information as up-to-date as possible.

The measurements themselves have been compiled from British and American surveys, which I have compared with findings of my own and those of some of my colleagues.

Metrication

It was decided to place metric and imperial measurements throughout the book. By 1972 the change to metric values and construction should be well established in industry. It is probable that the old imperial system will remain in use until the late 70s, and certainly until all sizes in imperial units are redefined in SI units.

General anatomical descriptive terms of position

The anatomical position
The anatomical terms of position assume that the body is standing upright and at 'attention' with the arms hanging by the sides with the palms of the hands facing to the front. The feet face directly forward. (Illustration **1a**.) This is a conventional attitude known as the 'anatomical position', and not a functional posture. The normal posture is shown in the illustration immediately following. (Illustration **1b**.) The median plane divides the body into two equal halves, and is an imaginary perpendicular centre section from the top of the head to the base of the figure between the feet.

Medial or *medial plane* means nearer, or a plane facing the median plane.

Lateral or *lateral plane* means farther away from, or a plane facing away from the median plane.

Anterior means facing the front, or to the front of a body part.

Posterior means facing the rear, or to the rear of a body part.

Superior means in the upper portion or nearer to the head.

Inferior means in the lower portion or nearer to the feet.

These six terms constitute pairs of relationship and used together enable an effective fix to be made on an anatomical point.

Superficial describes a position on or nearer to the skin surface.

Deep describes a position nearer the centre of an anatomical part or farther away from the skin. *Distal* and *proximal* indicate the relative ends of limb parts to one another, and they give positions in relation to the commencement of a limb. Thus the proximal end of the arm is the shoulder and the hand is at the distal end. However, by division down into smaller

1a and b

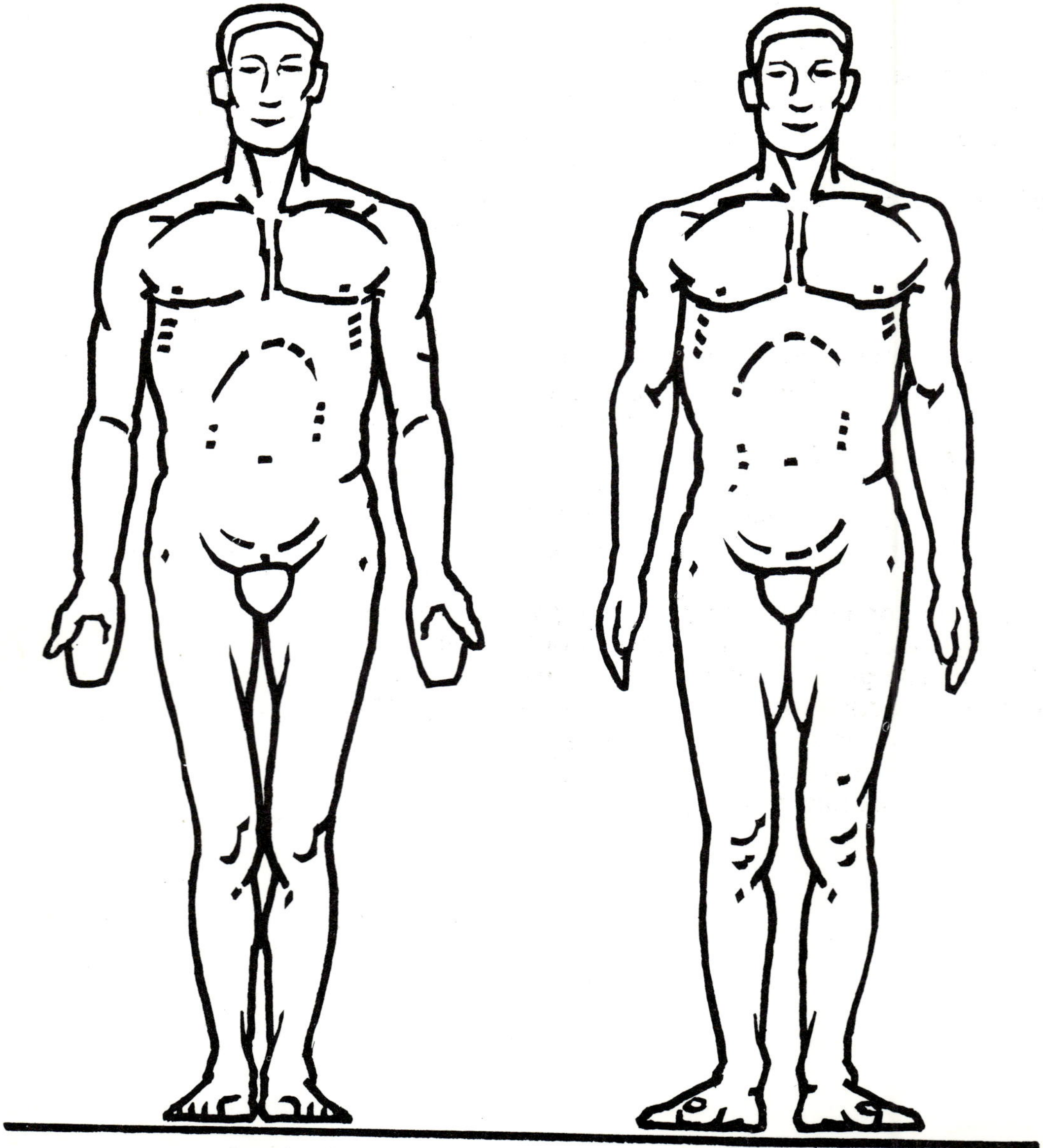

portions, the elbow becomes the proximal end of the forearm whilst the wrist is the distal end.

Three other imaginary planes or sections are also used. Any plane which is parallel to the median plane is called a *sagittal plane*, and any other perpendicular planes at right angles to these are called *coronal planes*. Any horizontal plane which would make a right angle to either a *sagittal* or *coronal plane*, or which would make a right angle to the long axis of a body part is called a *transverse plane* or *section*.

CHART 1 *Percentage weight of body parts*

Head	from	6% to 8%
Trunk	from	40% to 46%
Legs	for both	30% to 36%
Feet	for both	3% to 4%
Arms	for both	10% to 12%
Hands	for both	1% to 2%

There will be variation in the weight of body parts due to the variation in physique and sex. The extremity of the head however is generally greater in weight than the extremities of the hands and feet added together. The proportionate weight of the head in relation to the rest of the body parts should be noted because quite often care for head weight by a head rest is a necessary ingredient in a human situation.

The 'normal' or functional position for the body
The body is standing upright and at attention, although the bracing back of the shoulders and lift of the chest is 'comfortable' and not exaggerated to the extent of a parade ground manner. The arms hang freely by the side, the lower arm being pronated in such a manner that the thumbs face anteriorly. The feet are positioned slightly apart with the medial surfaces of the feet making an angle of about 30° to 45° between them. The angle being rather smaller during walking.

The line for the centre of gravity of the standing figure in the 'normal' position can be imagined pulling down perpendicularly through the three component masses: the skull, the thorax and abdomen – pelvis. These body parts making up something like 60% or more of body weight. If this pull of gravitational force is examined in side view it will be seen that it must pass through the legs and be contained inside the arch of the foot. (Illustration **1c**.) This is because, however varied man's structure may be, the centre of gravity must remain within its base, the feet. The normal standing position is in fact quite efficient; for the balancing trick that man is able to perform with the component parts of his body requires little muscular exertion. When standing restfully we relax our muscles but extend our legs and knees against the ground so that the ligaments of the joints brace them against collapse from the weight of the body above. The arches of the feet are retained by ligaments as well, although during lengthy standing foot and ankle joints may have to be reinforced by muscular lift and support.

The exact position of the centre of gravity in side view in the standing position is open to controversy. The variation of figure types makes it an impossible question to answer exactly.

From the arch of the foot upwards it seems that the centre of gravity must pass in front of the ankle joints and slightly in front of the knees. At the hips it will either pass through the joint itself or slightly above or behind it. Very rarely would it be found to lie in front of the hip joint, and then only in old age or infirmity. Its relationship to the vertebral column, the shoulder girdle and the head is very uncertain and depends greatly for its position in these regions on figure type and consequent posture. (Illustration **1c**.)

The normal posture in young children is different, in as much as the knees are not braced back so firmly, and the legs appear slightly bent.

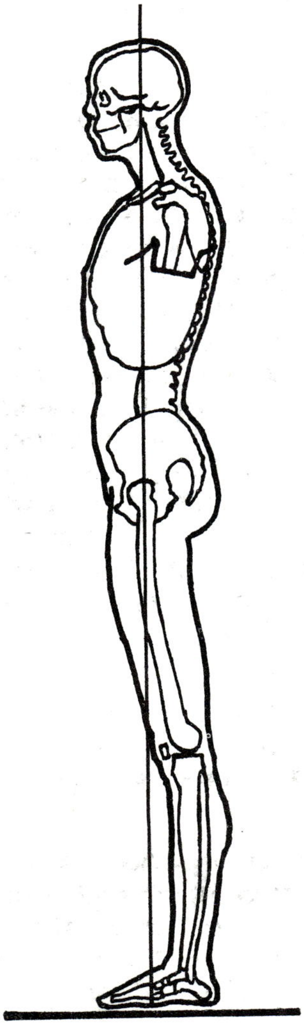

1c
To show normal posture side view. Here the line of gravity passes down through the mastoid process on the skull, through the acetablum in the pelvis slightly in front of the knee and then passes in front of the ankle joint. It must be remembered that this line of gravity is illustrated on a figure type that is highly meso-morphic. Note the four good curves to the spine. In less athletic postures the line of gravity might very well pass to the rear of the hip joint or acetablum.

2
As will be seen, human beings vary in proportions one from another. It is useful on occasions to have an average figure to refer to. In this illustration 'head lengths' have been used as a unit to obtain the varying bodily proportions found at different ages.

At one year a child is four heads high; at four years five

Section 1 **Growth, Maturity and Old Age**

Growth

During the whole developmental period between infancy and maturity many demands are made on the growing child by its environment. Although we may consider that much of the child's response to new situations is genetically programmed, their continuing growth and ability to survive

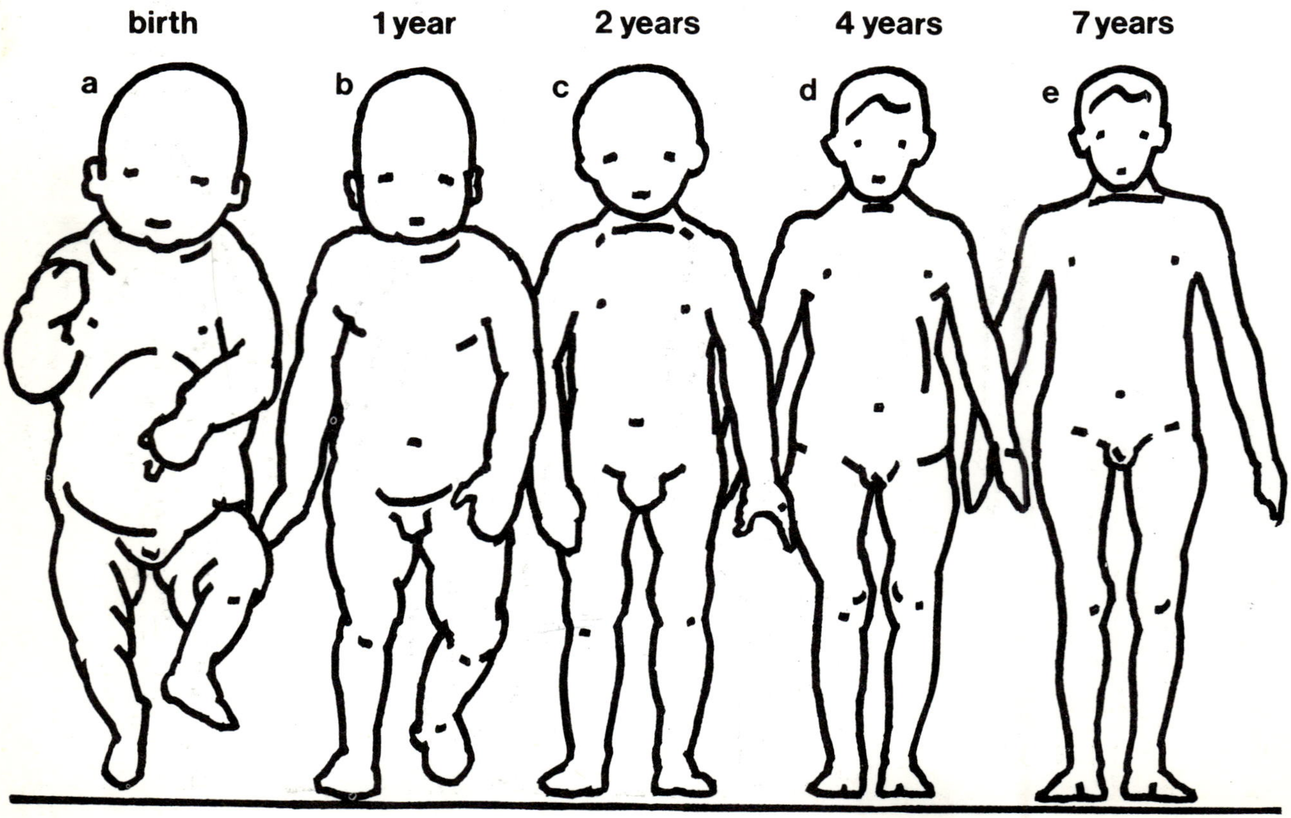

depends on a resilient and adaptable physiology working to sustain an ever greater mobility. This mobility is gradually attained by the extension and growth of the various organs and segments of the body. The growth rate of these organs and segments are not parallel but move at different speeds from each other.

heads, at nine years six heads; at sixteen years seven heads, and an adult is seven and a half heads. This is only a customary equivalent. A child at two years is half the height of an adult.

CHART 2 *Growth*

At birth	The arms are a similar length to the trunk
2 years old	The arms are 15% longer in length than the trunk
7 years old	The arms are 25% longer in length than the trunk
Adults	The arms are 50% longer in length than the trunk

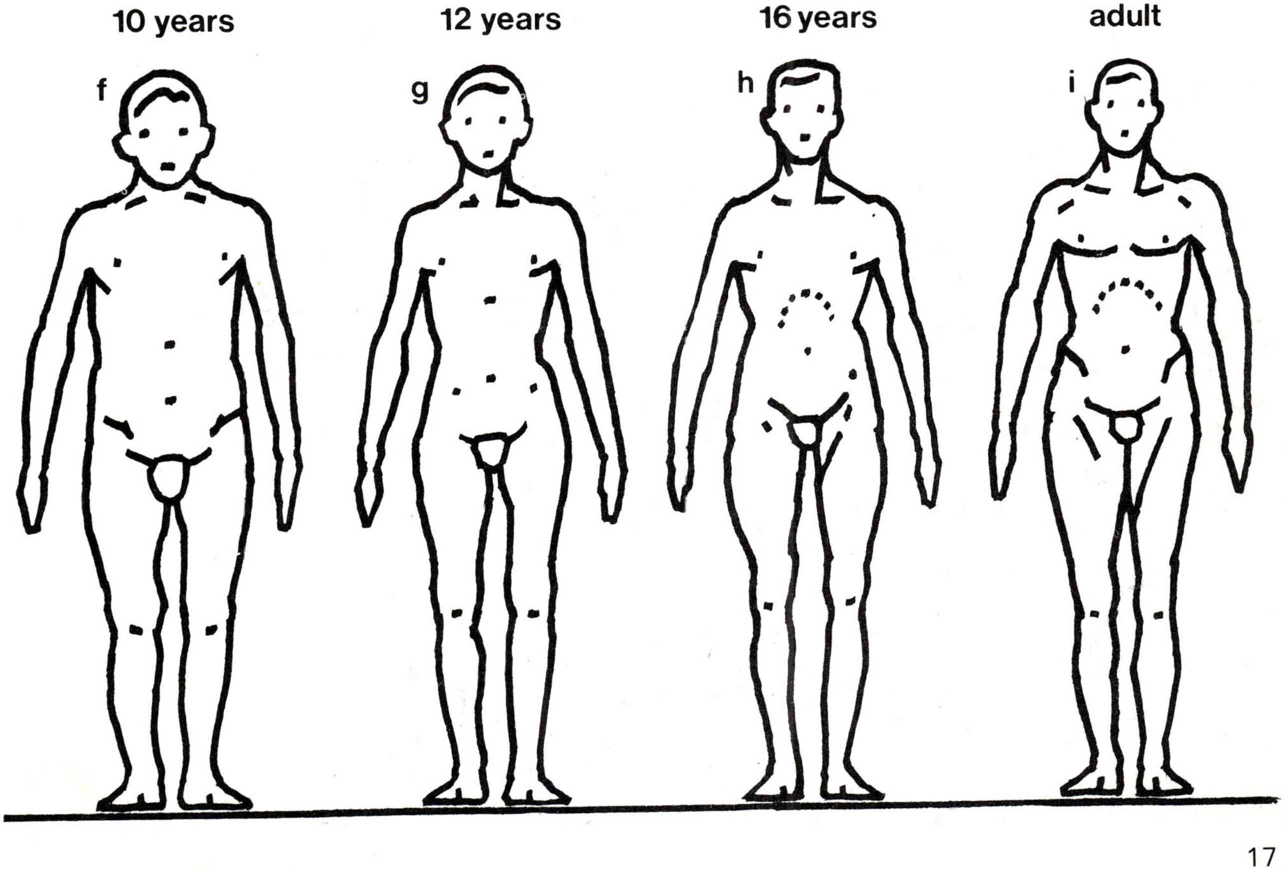

A growing child exhibits a gradual change in appearance and shape. (Illustration **2**.) Superficially it can be seen that some parts of the body are growing faster than others. A good example of this can be readily seen in the head, where it can be observed that for many years during early childhood the face remains relatively small in relation to the rapid increase of size of the cranium. (Illustration **3**.) The brain, so necessary for initiating action, approaches adult size quite early. It reaches 80% of adult size between four and five years of age: the time for joining school. The sexual reproductive system waits until later, when sometime about or after eleven years it makes fairly rapid growth and the difference between the sexes becomes steadily more manifest. The two different speeds with which these organs grow can be contrasted again with the increase in body weight. Body weight represents the sum total of all the body parts; and against time it can be informative about a child's progress towards maturity. Body weight shows a steady increase until about the sixth or seventh year of life when it is rather less than 50% of adult value. From seven to eleven years it almost marks time and so increases very little.

During the 'marking time' phase (of weight increase) glandular changes take place which are in the nature of a prelude to the later stages of growth in the skeleton and body tissues. From about eleven years onwards body weight again

3
Comparative development of child's face and head.

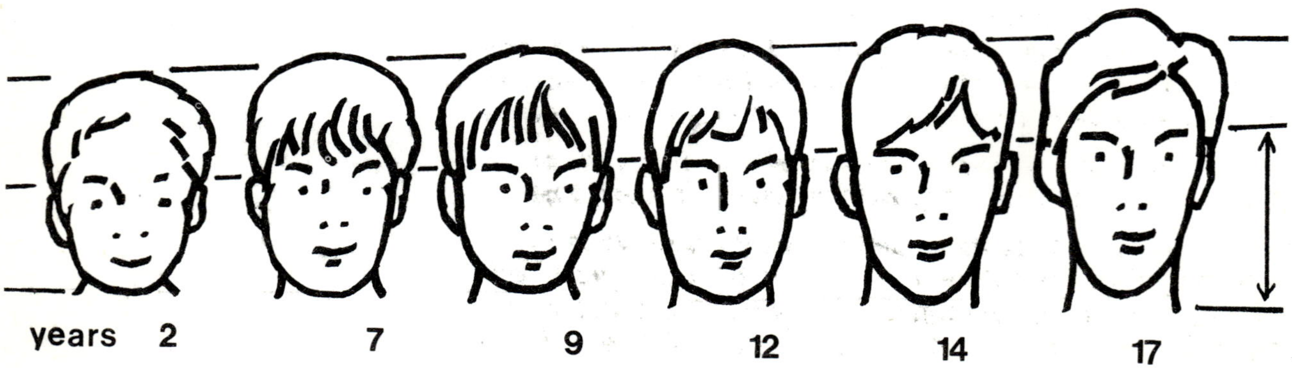

shows a regular increase until maturity. This increase is represented by a steady gain in standing height or statue; which in the first place is the result of the increase of leg growth.

It can be seen from the preceeding examples that there is often a wide variation of speed of growth between body parts, and always some variations while growth takes place. Externally this means that a child's body is continually changing its shape and therefore its proportions. So we can say that a child's body is liable and prone to change, and that these changes are normal and build up, stage by stage, towards a balanced mature condition.

The variation of the speed of growth between one body part and another is regulated so that each part reaches its size, proportional to the role it has to play in the body's physiology at the correct time. The endocrine glands provide the regulatory and 'time-clock' devices which control growth processes.

It is the alternation of the speed of growth between body parts that leads to a diversification and differentiation of one part of body structure from another. So it may be considered that the human body during development and growth changes in three ways: by increase in size, by differentiation of structure, and finally by the alteration of shape. These changes present three factors by which we may measure growth. The factors are three different types of dimensional increase: increase by length, increase by area and increase by weight. Between birth and maturity the approximate increase overall in each of the three dimensions is: height, an increase of three and a half times: skin area, an increase of seven times: weight, an increase of twenty times.

Although these increases are dissimilar from one another there is a typical mode of growth increment in each case.

The increase of these three factors would give a curve of different proportions if the values of the increases were each plotted against time. All three curves would be asymmetrical but they would have one thing in common; each would have a single highest point or peak, and the peak would indicate the time of the greatest increase in that particular dimension. As the curves were different the peaks would not coincide. These unmatchable curves illustrate the variety of the rates of progress in different dimensions.

It is because of the variety of the rates of progress in different dimensions during the growth of a child, that it is best to consider it to be a process of development as well as growth.

The human being has the most protracted period of growth and development of any animal. Man's childhood, for instance, is about as long as the total life expectancy of a monkey. An average of 20 years is required for man to develop to full maturity. Twenty years is given here as a 'safe'

4
Three girls of 11·5 years.

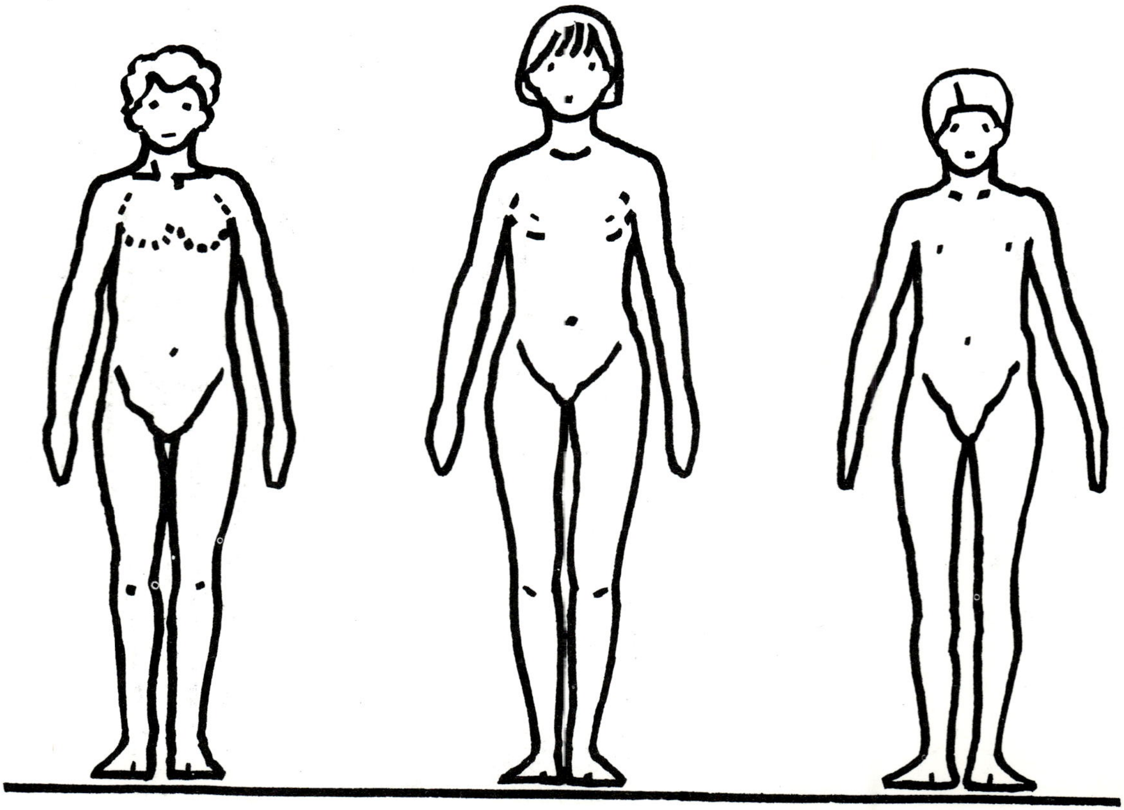

figure for physical maturity. The numerical age of a child cannot be taken as a point of reference when establishing its progress towards physical maturity. To say that a girl is 13 is not any evidence of her having reached a particular stage in puberty. Nor is it any evidence of her size in any bodily dimension. Numerical age is less important than body size and proportion achieved.

When studying physical development, and leaving numerical age aside, we still meet difficulties.* With young children often the short ones grow more slowly; yet in adolescence the taller ones grow more slowly, and the shorter ones grow

* Tanner, J. M. *Growth at Adolescence,* Blackwell Scientific Publications, 1962.

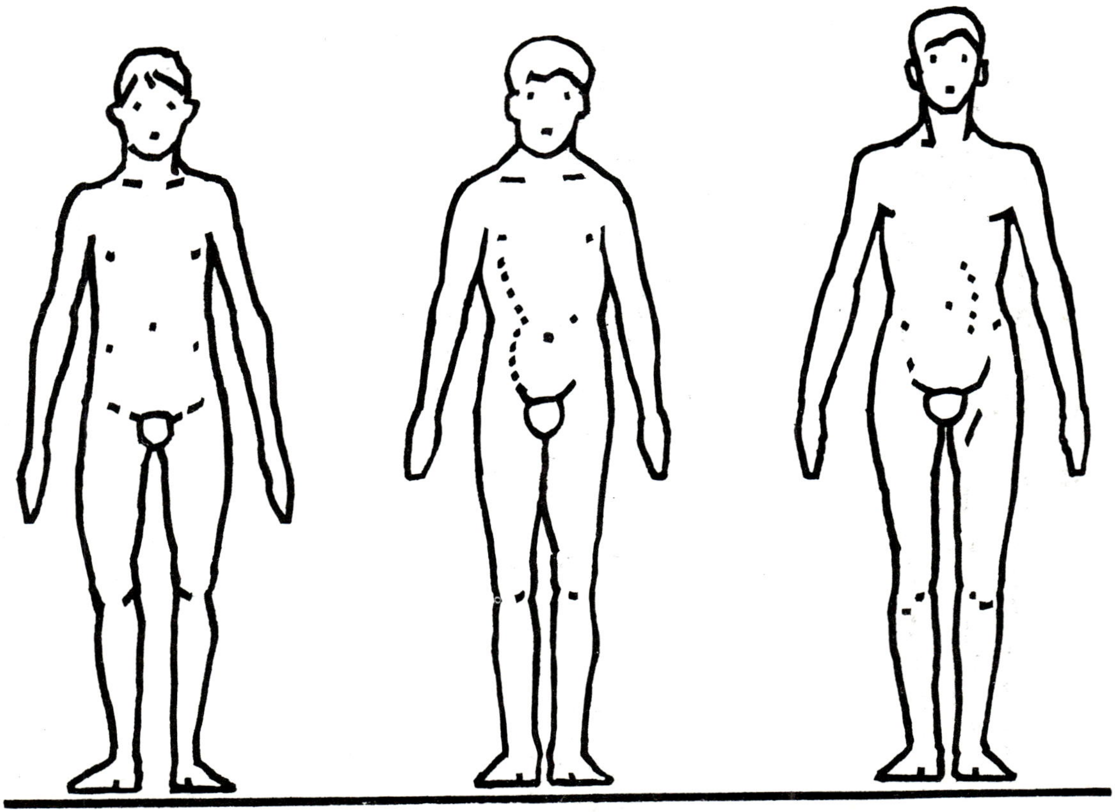

5
Three boys of 12·5 years.

CHART 3 *Boys: Some typical length and girth measurement relationships found from age five to fifteen in inches and millimetres*

Years	Arm length		Leg length (*Crotch to ground*)		Chest girth		Waist girth		Hip girth	
	in	mm	in	mm	in	mm	in	mm	in	mm
5	17·5	444	19	483	21	533	20	508	21·5	546
6	18·5	470	20	508	21·5	546	20	508	22	559
7	20	508	21·5	546	22	559	20·5	521	23	584
8	21	533	22·5	572	22	559	20·5	521	23	584
9	22	559	24	610	24	610	21·5	546	25	635
10	23	584	25	635	25	635	22	559	26	660
11	24	610	26	660	25·5	648	23	584	27	686
12	25·5	648	27	686	26·5	673	23·5	597	28	711
13	26·5	673	28·5	724	27·5	698	24	610	29·5	749
14	27·5	698	30	762	29	737	25	635	31	787
15	29	737	31	787	30	762	26	660	32·5	826

CHART 4 *Girls: Some typical length and girth measurement relationships found from age five to fifteen in inches and millimetres*

Years	Arm length		Leg length (*Crotch to ground*)		Chest girth		Waist girth		Hip girth	
	in	mm	in	mm	in	mm	in	mm	in	mm
5	17·5	444	19	483	21	533	19·5	495	22	559
6	18·5	470	20	508	21	533	20	508	22·5	572
7	20	508	21·5	546	22	559	20	508	23·5	597
8	21	533	22·5	572	22·5	572	20·5	521	24·5	622
9	22	559	24	610	23·5	597	21	533	25·5	648
10	23·5	597	25	635	25	635	22	559	27	686
11	24	610	26	660	25·5	648	22	559	28	711
12	25	635	27·5	698	27	686	23	584	29·5	749
13	26	660	28	711	29	737	24	610	32	813
14	27	686	28·5	724	30	762	25	635	34	864
15	27·5	698	29	737	31·5	800	25·5	648	35	889

more rapidly for a longer period of time. These type of changes in growth rates tend to accentuate the great variability of children's body measurements. For any numerical age, in either sex, we are likely to be presented with a wide range of body builds. (Illustration **4** and **5** and Charts **3** and **4**.)

When considering the physical growth of the human body in detail we find that although there may be extension or growth in one segment in a transverse direction, in the immediate adjoining segment growth may be taking place in a longitudinal direction. Growth of body parts can be seen as an example of the biological law of alternation of growth. During growth the bones of the skeleton undergo necessary changes of proportion by surface accretion accompanied by internal erosion. The law of alternation of growth can be seen to be clearly operating in the change of proportion taking place during the growth of the long bones of the skeleton. It can be considered that the long bones thicken for a period ranging from four to six months and then lengthen during the next four to six months. For example the femur bone of the leg thickens while the tibia and fibula bones lengthen. Then in the next four to six month period the femur lengthens while the tibia and fibula thicken. Throughout the growing period there is an extensive remodelling of all the bones of the skeleton taking place in this manner. This remodelling is paced by growth in the superficial tissues, and as the internal structure undergoes some corresponding alteration takes place in the external forms of the body. A child's figure is not like an adult's in miniature. It re-forms and almost seems to deform, as it progresses. The same number of parts of an adult are present, but they are continually being presented in different proportions. As the law of alternation of growth suggests there are times; when feet seem too big for legs; when hands seem to be too big for arms; or lower arms for upper arms; and these instances can be easily multiplied. (Illustration **6**.)

In early infancy there is a general chubbiness, with relatively large dimensions of the trunk and head. (Illustrations

2a and **2b**.) The upper part of the body is proportionately larger than the lower. The head length and breadth and circumference are about 96% of the adult value by ten years of age.

There is a great difference in the shape of the thorax between the newborn and the adult. In the first year of life the horizontal section of the thorax is cylindrical, the lower ribs are horizontal and the upper part appears constricted. (Illustration **2b**.) By the second year the horizontal section

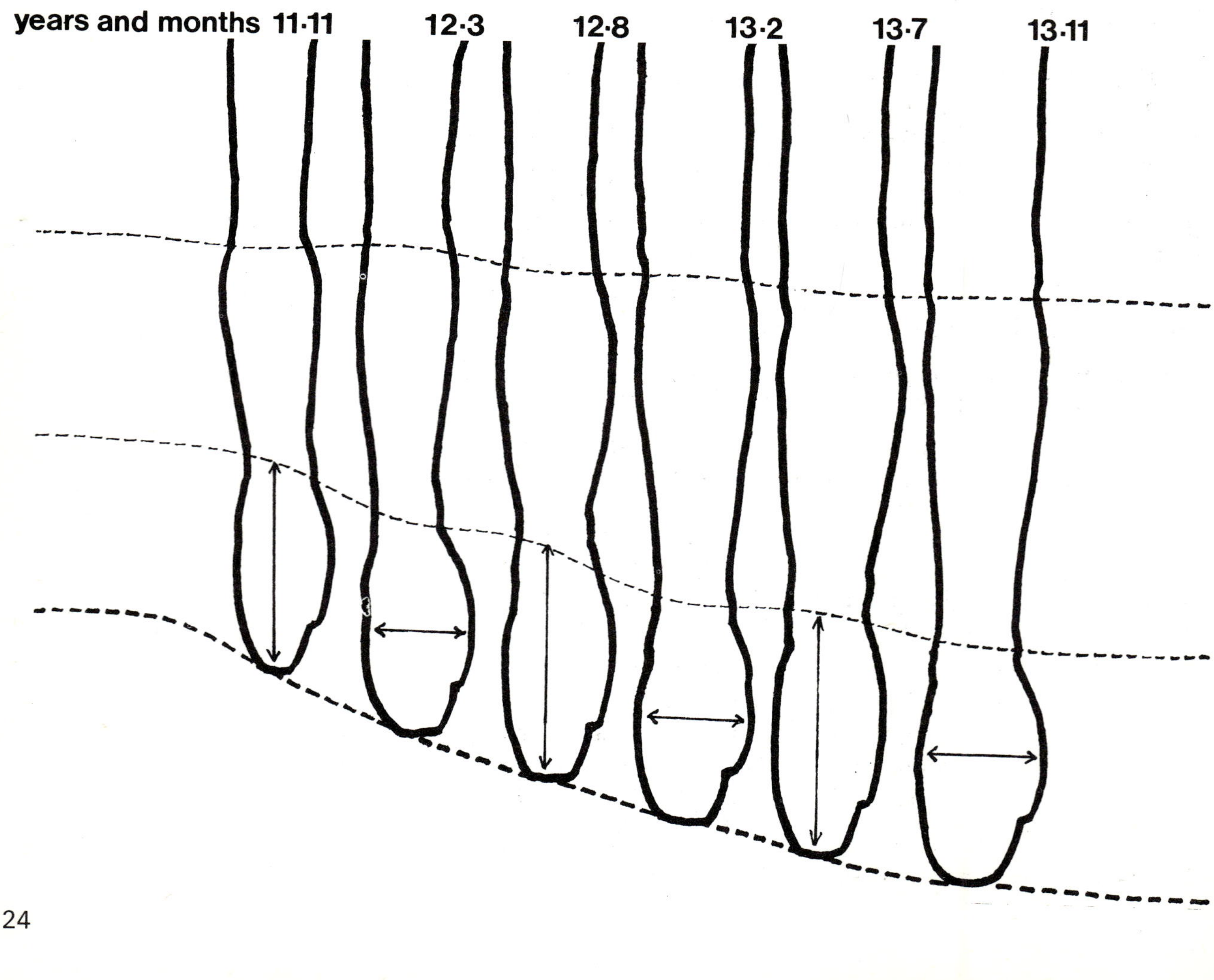

6
The consideration of the growth of a child's limb in segments which are growing alternately by length and by transverse section. This example is based on the possible development of a girl's arm.

becomes oval. The vertical thoracic dimensions are shorter in infancy and the thorax appears very small compared with the abdomen. (Illustration **2b**.) The new-born are, in fact, all digestion or indigestion without means of locomotion. Until about six years the limbs appear relatively short and puny. After six, during the period of juvenile growth, the legs and arms increase in relation to the trunk. (Illustration **2e**, **f** and **g**.) Throughout childhood the speed of growth for arms and legs remains different from the speed of growth for the trunk and head. With the growth of the limbs the growing child begins to fulfil the fuller role of the upright primate. Maintaining their upright posture with greater ease and developing an ever freer use of the fore-limbs for grasping, swinging and manipulation, and also showing an ever-increasing tactile appreciation with the hands. At a later stage of development, with a larger chest and longer clavicles and greater power to pronate and supinate their fore-arms, the growing child can obtain a greater vigour of the arm and an increased dexterity of the hand. (Illustration **2g** and **h**.)

Until the end of the juvenile stage at about nine years the growth for both sexes is much the same.

During pre-adolescent years growth is largely concerned with building up a strong skeleton and extending tissues.

Between ten to eleven years, or the pre-puberal period, more height is continually being gained by the continuing growth of the lower limbs. (Illustration **2f** and **g**.) At about 12 years the cervical vertebrae lengthen and the neck becomes apparent for the first time. (Illustration **2g**.) The clavicles begin a descent to the horizontal which is their position in adult life. During this time the trunk retains a small appearance proportionately to the whole statue.

The waist, however, is becoming more defined and greater movement is possible between the pelvis and the thorax. At about this age the external physical differences between the sexes begin to make their appearance as a result of the processes of sexual morphological differentation.

The period of puberty comes at different times for boys and

girls; girls being something like two years ahead of boys. Girls are generally heavier, taller and have a greater body surface area than boys in the early pre-puberal period, but are then generally lighter than boys to puberty. Boys begin to assume extra weight in the later pre-puberal period.

The segments of the appendages continue growing, and in particular the proportion of leg length to trunk length increases through early and middle adolescence. (Illustration **2h**.) The thorax remains narrow, although the slope of the shoulders is beginning to increase because of the extra weight being gained in the arms.

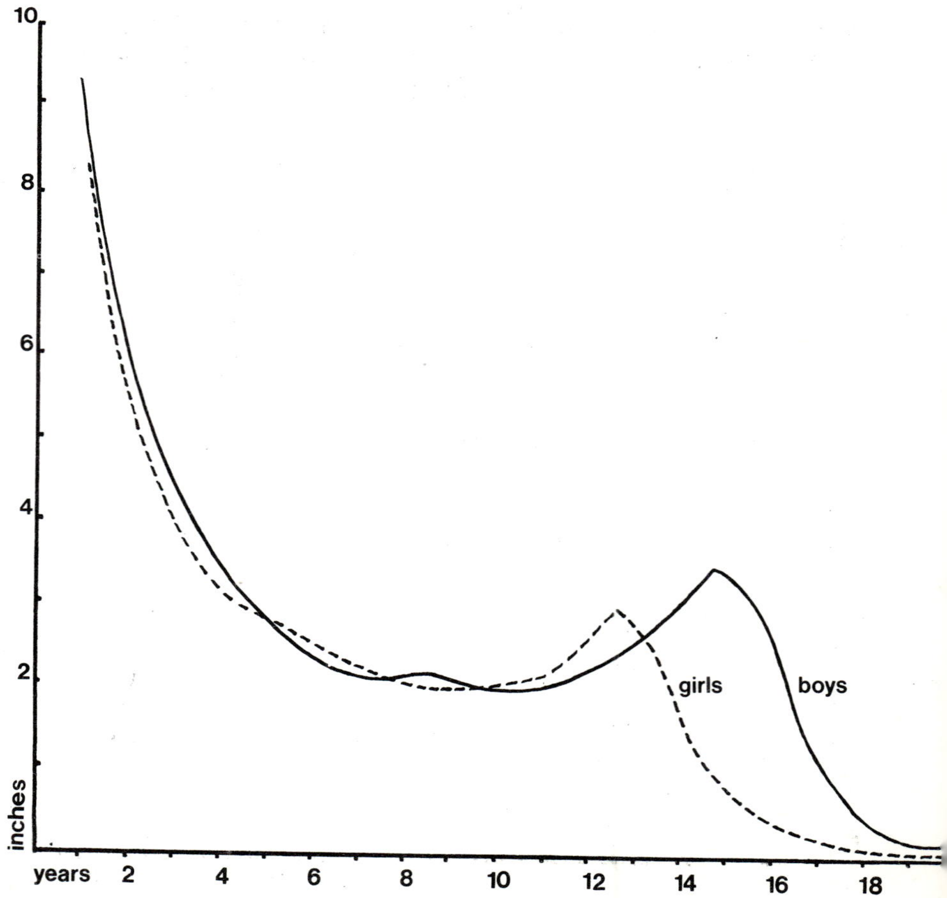

7a
Graph lines depicting comparative increases of girls' and boys' heights in inches each year. Upward peaks in teens are the adolescent growth spurt. The boys' growth spurt can be seen to be later and greater, giving boys extra increments for standing height and also adding larger body dimensions generally.

From about the 11th to the 13th year in girls and from about $12\frac{1}{2}$ to $15\frac{1}{2}$ year in boys there occurs a remarkable acceleration of growth. (Illustration **7a**.) This acceleration is clearly noticeable visually, and is very strongly marked if data of one dimension of a growing child is plotted in graph form. (Illustration **7b**.) During the period of the adolescent growth spurt there is a very marked increase in size of various body parts. The spurt of growth starts with an increase in foot length. In general growth spurts commence in the distal segments of the appendicular skeleton and proceeds through to more proximal ones. In the leg region, after the growth of foot length, the lower leg extends in length, and this is followed by the extension of the upper leg and thigh. At all ages of childhood the feet dimensions keep ahead of arm, leg and trunk dimensions and are nearer adult size. Increased spurts of growth in the feet may come a year to two years

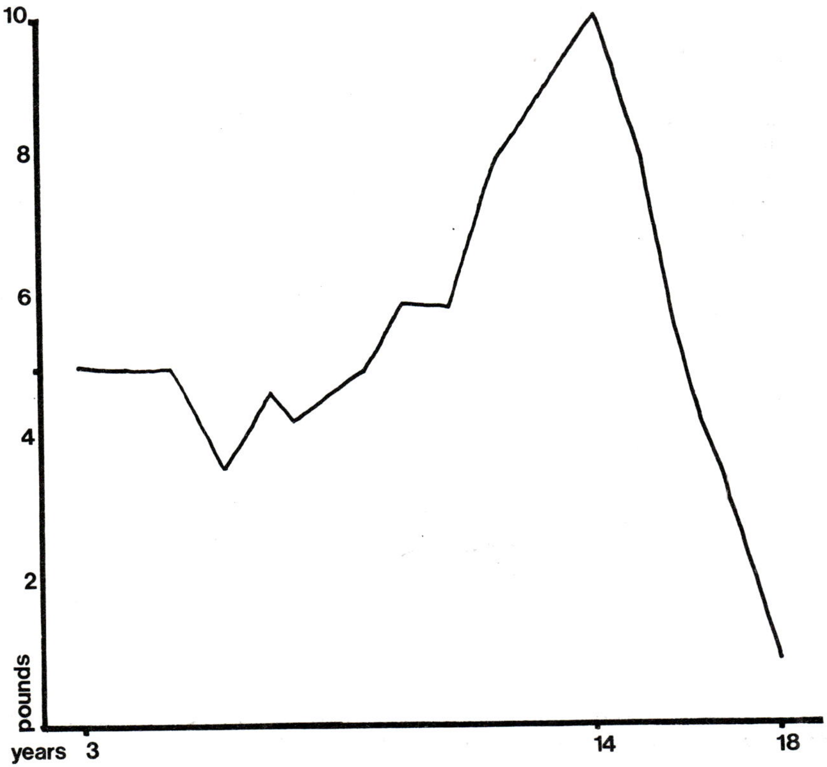

7b
Incremental increases in pounds per year for a girl from three years to 18 years.

Maximum increase in this case was at 14 years. Child's final weight was about 110 lbs (49·89 kg).

ahead of those of statue. Exceptionally rapid growth in the feet is exclusive to girls; and for those girls who have attained marked increase of dimensions in this region early, subsequently their feet will then cease to grow much more. Maximum velocity of foot growth can be reached at ten years of age for girls, whereas a boy's maximum is at about 12 or 13 years. Velocities of growth vary throughout post-natal development, and the increase of foot length growth has periods of acceleration. However, final foot growth probably takes place in breadth dimensions and not in its length.

Three to four months after the feet begin to increase the hip and chest breadths commence their growth spurt. Boys' shoulders and trunk height increase in relation to hip width and the definition of male proportions commences. (Illustration **2h**.)

CHART 5 *Variation in the range of boys' and girls' heights from birth to maturity.* The range is from the 5th to the 95th percentile

	Range	Difference
Birth to six years	± 1·25 inches (32 mm)	2·5 inches (64 mm)
From six to eleven years	± ·9 inches (22·5 mm)	1·8 inches (45 mm)
From eleven to fourteen years	± 1·25 inches (32 mm)	2·5 inches (64 mm)
From fourteen years to maturity	From ± 1·25 inches (32 mm) at fourteen years to ± 4·25 inches (108 mm) at maturity	2·5 inches (64 mm) to 8·5 inches (216 mm)

Range of heights for each age group for either sex increases with age, although ranges of height do not appear quite so great between six and twelve years old. After 14 or 15 years of age until maturity the range of heights increases to an extreme of ± 8 inches (203·2 mm) each side of the mean value. 5th and 95th percentiles giving values of ± 4 inches (101·6 mm). The whole range of normal adult height can be found between 55 inches (1397 mm) and 79 inches (2006·6 mm).

Girls on the other hand have a spurt in hip-width growth and so begin to develop the typical female silhouette. Sexual differences of proportion become more marked in the trunk region with the development of breasts by the female. (Illustration **9**. Compare charts **3** and **4**.) Also with a less broad thorax and shoulder girdle, and a slightly lower sternum bone than boys, a girl's neck and shoulder regions begin to show their more characteristic delicate structure.

Leaving aside the development of mammary glands and breast tissue, boys and girls have a fairly similar increase rate for chest development. In the head the face has remained short during infancy but as the last of the permanent teeth erupt a spurt of growth takes place in the facial bones. The face becomes longer and the chin more prominent, and the more purposeful look of the adult makes its appearance.

In the arm region the sequence during the growth spurt follows a similar order to that of the leg. That is, the spurt of growth starts in the distal segment of the hand and by alternating four- to six-month periods progresses to the upper arm. The only distinction between the longitudinal growth of the skeletal parts of the arm and leg is, that whereas the growth of the bones of the arm takes place mainly at the proximal end of the humerus and the distal ends of the ulna and radius, this order is exactly reversed for the long bones of the leg; the femur bone growing mainly at its distal end, and the tibia and fibula growing at their proximal ends. It would appear that these main growing ends have been selected in the course of man's evolution, so that the joints of the elbow, hips and ankle are prepared early in a child's development to withstand the stress and strain of his erect position and the practice of his physical functions. (Illustration 8**b** and **c**.)

In early and middle adolescence it is the leg growth that takes the entire body towards mature height. Mature proportion and final height is obtained in late adolescence and sub-adulthood by growth of the trunk. Leg height and trunk and head height are the two constituents of stature. (See Chart **6** for typical heights.)

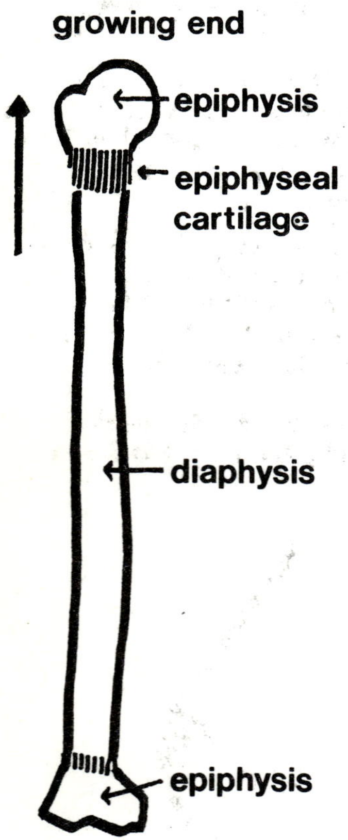

a

8
Growing ends

Growth continues for the greatest time at a preferred end of a bone. The growing ends of the arm and leg bones are illustrated here in (**b**) and (**c**). (**a**) illustrates diagramatically the portions of a growing bone. The central section or primary centre is the diaphysis and the secondary centres at the ends are called the epiphysis. An epiphyseal section of cartilage separates the two centres and it is its constant adaptation to the processes of ossification taking place on its borders that facilitates bone-length growth. Finally, at about the time of puberty the primary and secondary centres of ossification unite and the proportions of the adult bone begins to be laid down.

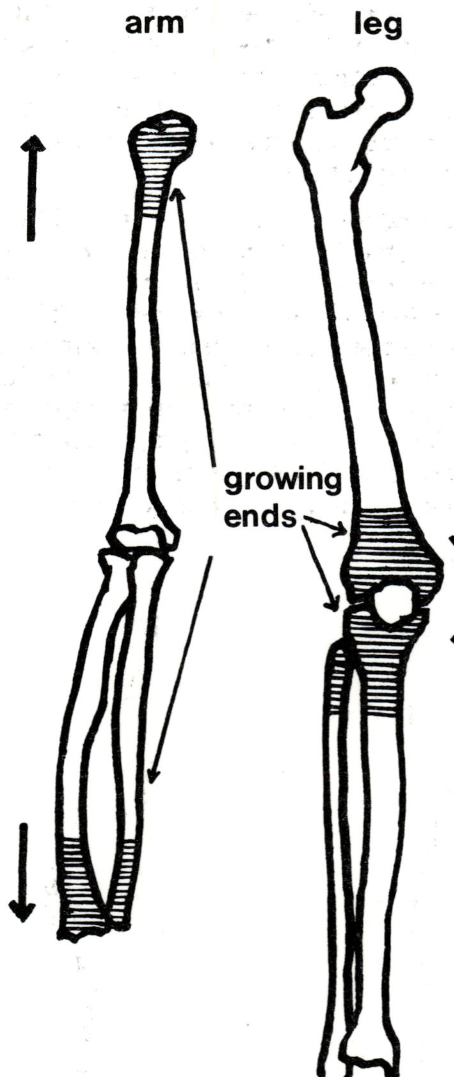

b c

Growth does not cease abruptly, but tails off finally in the regions of the trunk and the head. The trunk may continue to increase in height until 25 years of age or more, and the head measurements may continue to increase, in a very small percentage of cases, for a much longer period. Forms of late growth can be confusing as there is a possibility that one may be observing pathological over-growth. These are restricted individual concerns and lie outside the designer's appraisal of growth.

We find the most obvious sex differences in body physique illustrated by changes of body size and shape. Differential growth rates at adolescence in the sexes vary as regards shoulder and hip breadths, trunk heights, head size and the proportions of arm and leg segments. (Illustration **10**.) The adolescent growth spurt occurs later in the male than the female, and continues for a longer period. The overall difference in size and proportion between adult men and women is largely the result of the adolescent spurt of growth. This longer growth period for men gives them larger bones

9
Breast development

(a) Breast bud. The commencement of the body of the breast.

(b) The enlargement of the breast body and the definition of the areolae as separate eminences.

(c) Continuing enlargement of the breast body and the separate and distinct elevation of both the areolae and the papillae.

(d) Adult stage. The attainment of the conical or hemispherical form of the mature breast body with the papillae fully elevated.

In the diagrams the areolae are indicated by shaded areas.

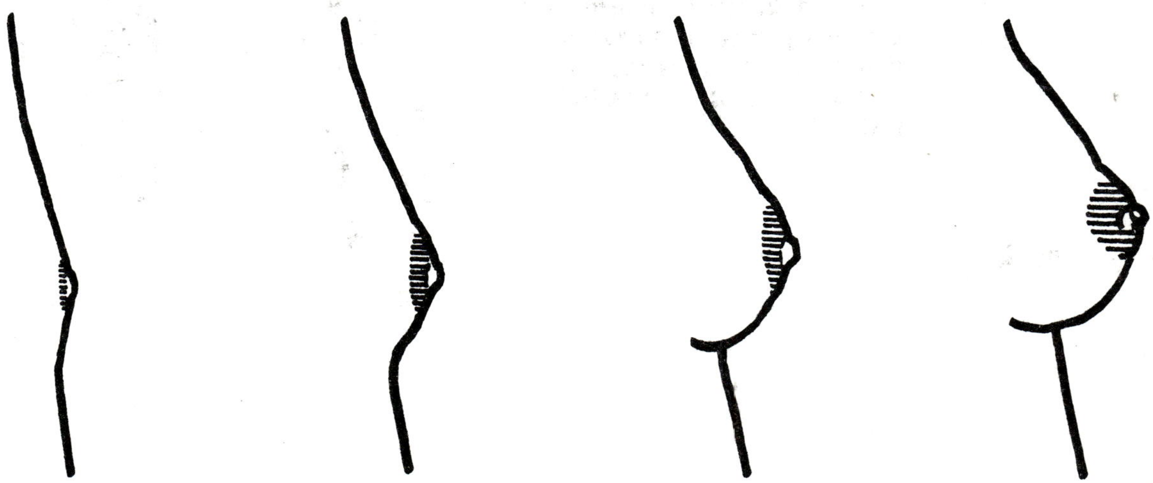

a b c d

and a greater body size. It also adds extra increments to standing height. (Illustration **7a**.) Boys from birth are averagely bigger at all stages of growth except during the early part of the adolescent growth spurt.

Boys' and girls' strength and endurance increase with the growth of bone and muscle. Generally the peak of strength comes about a year after the achievement of the peak of height.

10

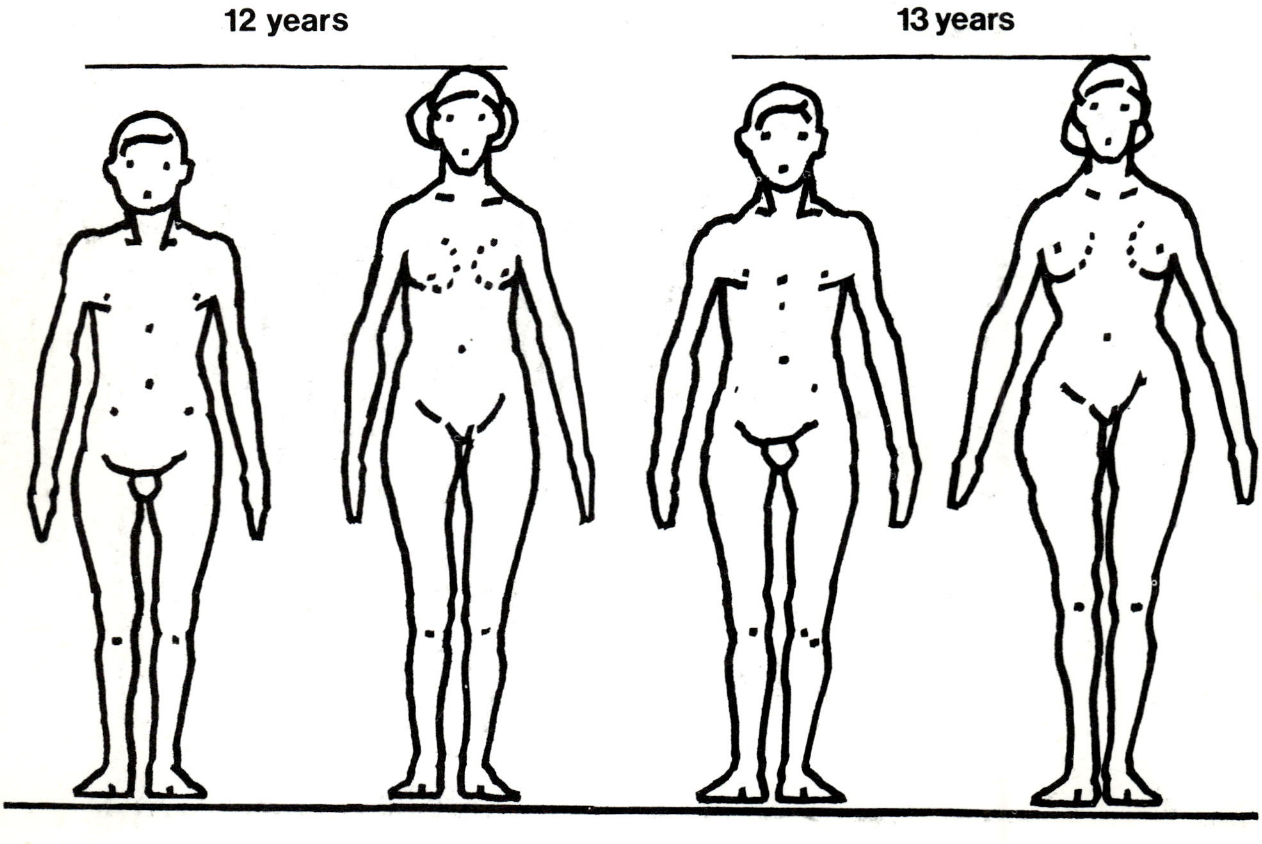

Although boys wait until a later age than girls to gain maximum strength it is then superior to girls due to the larger muscles and bones they have built up. The greater strength achieved by the male is proportionate to the greater spurt of growth at adolescence which creates larger bones and muscles. Greater strength in boys and men is also the result of the broader shoulders of their sex, which is a noticeable feature of male sexual dimorphism as the broader pelvis is a feature of the female sex.

10

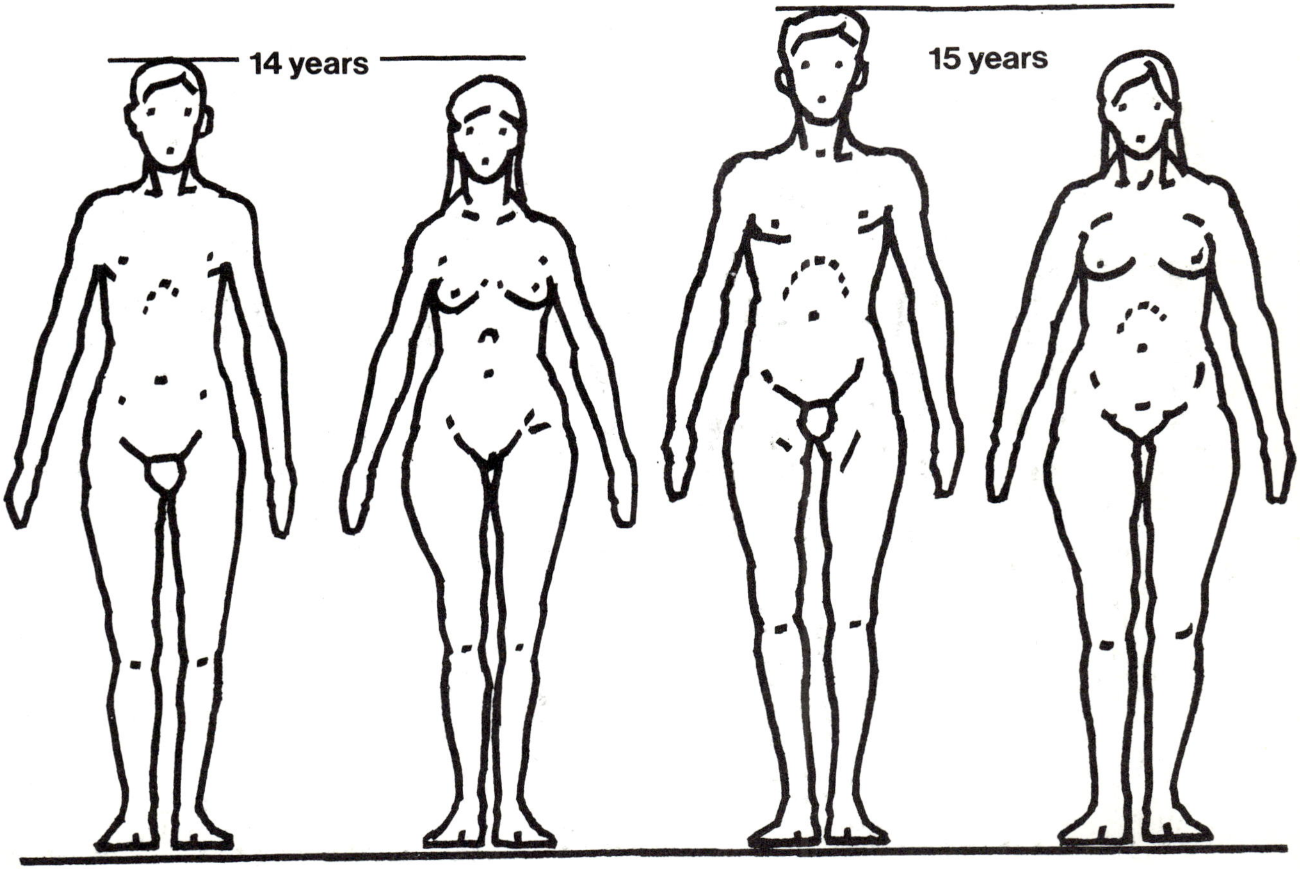

Maturity

In general the differences between the sexes in the skeleton are as follows: All the eminences of the male skull are more angular and nearer to 90° than the female; the only exception being the temporal eminences, which are more marked with the female than with the male.

CHART 6 *Typical heights for both sexes from 1 to 19 years of age in inches and millimetres*

	Male		Female			Male		Female	
	in	mm	in	mm		in	mm	in	mm
Birth	19·9	505·5	19·8	503	11 yrs.	56·8	1442	57·3	1455·5
1 yr.	29·6	752	29·3	744·5	12 ,,	58·3	1480·5	59·6	1514
2 ,,	34·5	876	34·3	871·5	13 ,,	60·6	1539	61·3	1556·5
3 ,,	37·8	960	37·5	952	14 ,,	63·8	1620	62·7	1592·5
4 ,,	40·7	1033·5	40·6	1031	15 ,,	66·3	1683·5	63·5	1613
5 ,,	43·7	1109·5	43·7	1109·5	16 ,,	67·7	1719·5	63·8	1620
6 ,,	46	1168	45·5	1156	17 ,,	68·2	1732	63·9	1622·5
7 ,,	48·2	1224	48	1219	18 ,,	68·4	1737	64·0	1626
8 ,,	50·5	1283	50·3	1226·5	19 ,,	68·6	1742	64·1	1628·5
9 ,,	52·6	1336	52·3	1328·5					
10 ,,	54·5	1384	54·7	1389·5					

At two years approximately half of adult height has been achieved.

The male trunk shows a broader and longer thorax with longer clavicles and greater scapulas. The male neck appears shorter because of the higher position of the manubrium in relation to the vertebrae prominens. The female pelvis is wider and altogether more spacious. The sacrum is much nearer the horizontal and there is a much smaller lumbar-sacral angle. This gives the female a more mesomorphic appearance at the pelvis in side view. (Illustration **19**.) The area of the ilium is less but the pubic arch is lower and wider, and the acetabulum face more anteriorly and inferiorly. The greater width of the base of the female pelvis means that their pubic-femur relationship is nearer an equilateral triangle. (Illustration **12**.) The bones of the legs and feet of the male are longer than the female. The bones of the arms and hands are also longer, but this is less obvious visually than the difference in the legs. 85% of females can extend the elbow joint to beyond 180°, and have a proportionately longer humerus bone in relation to the radius and ulna bones. (Illustration **11**)

CHART 7 *Possible decline in stature in inches and millimetres*

Years	Male		Female	
	in	mm	in	mm
20–24	None	None	None	None
25–29	None	None	·2	5
30–34	·2	5	·1	2·5
35–39	·2	5	·1	2·5
40–49	·4	10	·2	5
50–59	·5	12·5	·4	10
60–69	·6	15	·6	15
70–79	·3	7·5	·6	15
80–89	·3	7·5	None	None

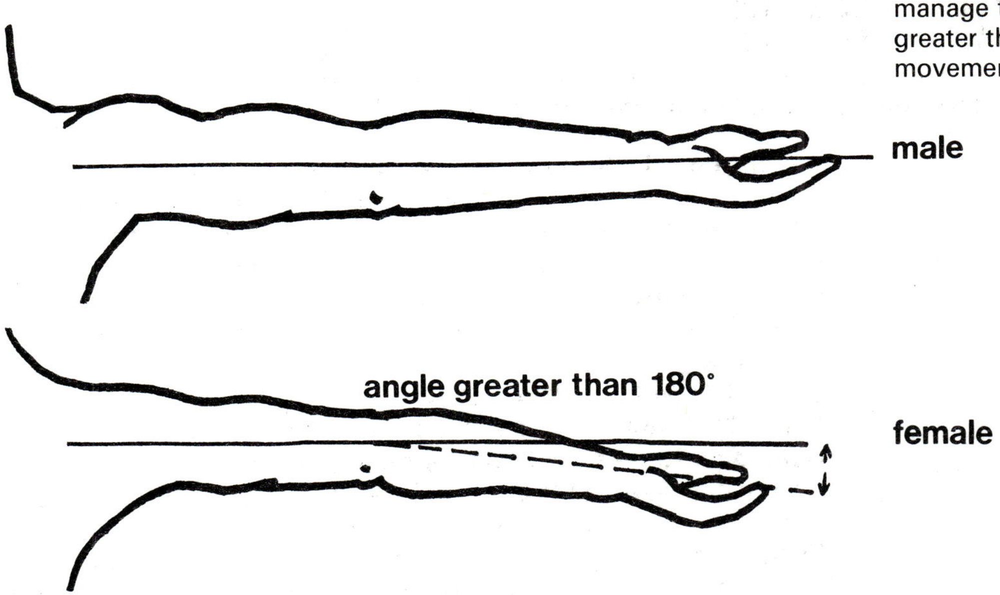

11
Elbow extension, anterior view.

Male maximum 180°.

Female a percentage can manage to produce an angle greater than an 180° with this movement.

Some of the more obvious differences in the skeleton between the sexes may be summed up as follows: the male has wider shoulders and a greater thorax, and a relatively narrow pelvis; a larger skull and longer arms and legs with bigger hands and feet. The female has narrow shoulders and a rounder and

12
Femur-pelvic relationship.

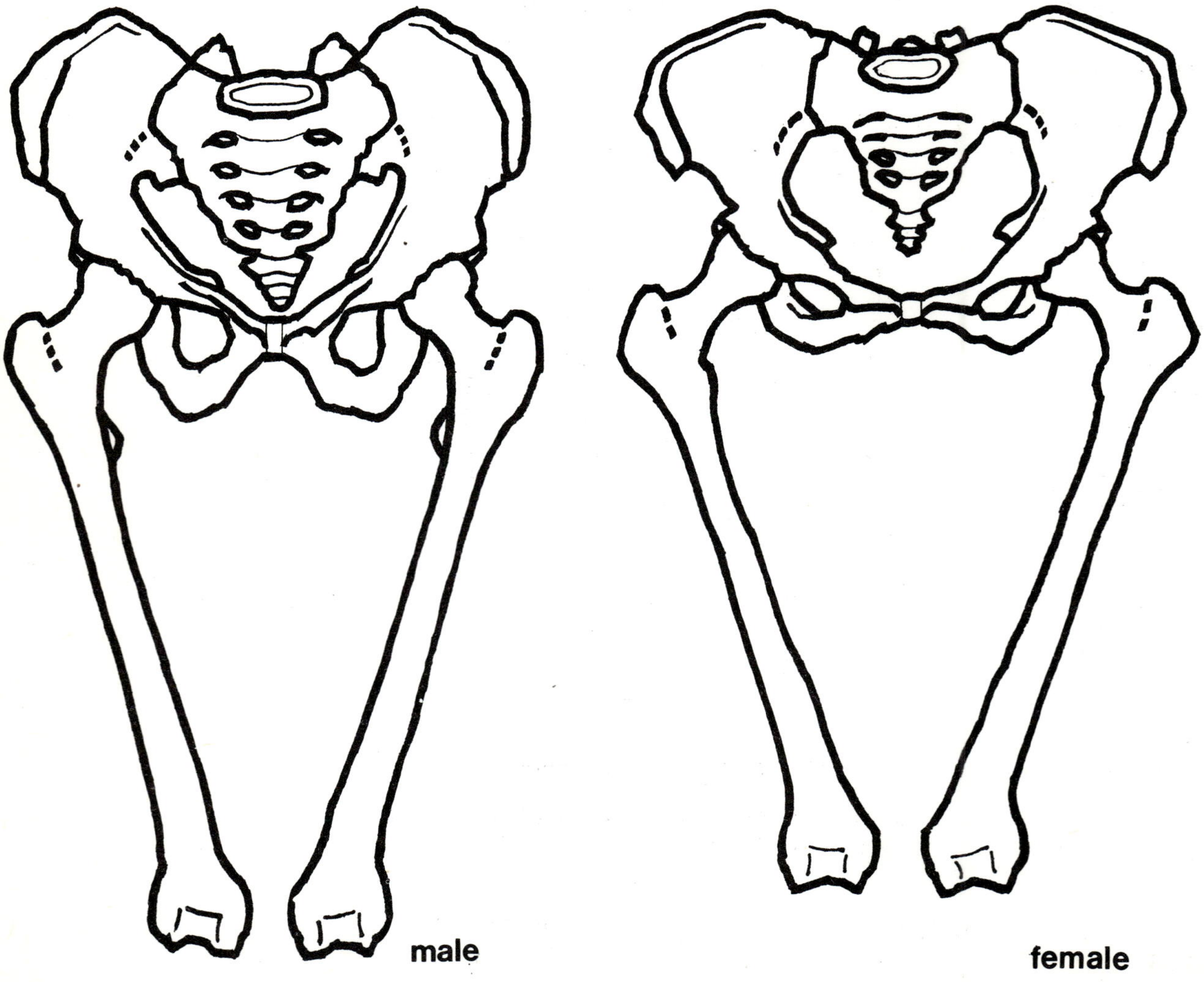

slighter thorax with a relatively wider pelvis which is tilted more anteriorly; a smaller and more rounded skull, shorter arms and legs with smaller hands and feet. The average difference between men and women's heights is about six or seven per cent.

The final picture of the male and female form is made up by differences between superficial body tissues. The male figure has more muscle to fat than the female. The average male will have a ratio of six to three of muscle to fat; the average female, a ratio of five to four of muscle to fat.

CHART 8 *Typical weights for children, men and women*

	Boys & Men		Girls & Women	
	lbs	kgs	lbs	kgs
2 yrs.	28	12·70	27	12·25
4 yrs.	36	16·33	35	15·88
6 yrs.	45	20·42	41	18·60
8 yrs.	55	24·95	55	24·95
10 yrs.	70	31·75	70	31·75
12 yrs.	84	38·10	88	39·92
15 yrs.	124	56·25	115	52·16
19 to 20 yrs.	156	70·76	123	55·79
20 to 30 yrs.	162	73·48	130	58·97
31 to 40 yrs.	166	75·30	134	60·78
41 to 50 yrs.	169	76·66	145	65·77
51 to 60 yrs.	166	75·30	149	67·59
61 to 70 yrs.	163	73·94	147	66·68
71 to 80 yrs.	152	68·95	140	63·50

Weights

Only ten percent are over 200 lbs (90·72 kg) in weight.

Typical women's weights may range from 5tn percentile value of 100 lbs (45·36 kg) to 95th percentile of 176 lbs (79·83 kg).

Typical men's weights may range from 5th percentile value of 126 lbs (57·15 kg) to 95th percentile of 213 lbs (96·61 kg).

From these ratios it is at once evident that there is a predominance of body fat in the female. Girls also have more subcutaneous fat than boys from birth, more noticeably so from about seven years onwards.

With the rapid expansion of the transverse sections of the body segments during the adolescent growth spurt, girls' growth is more in fat than muscle, though it may be a less apparent accumulation in the arms. In the trunk and upper thighs the acquisition of fat by growing girls can be quite large. Boys show varying amounts of body fat up to adolescence, and then with an increase of bone and muscle growth to fat, give the appearance of the 'lean and hungry youth'. A boy's growth spurt is in bone and muscle, when his larger and more powerful physique is being assumed. Boys growing in muscle and remaining stationary in fat will actually appear to loose fat. After the adolescent growth spurt a proportion of boys will assume extra fat again as their bodily constitution consolidates its mature form.

The manifestation of particular ratios of bone to muscle, and muscle to fat during adolescence is the determinant of sex in physique.

Subcutaneous fat in the normal adult has a predilection for residence in particular body areas. This is so for both sexes, although its flowering abundance in certain areas is reserved for the female during sub-adulthood and maturity. The reason for the residual areas of subcutaneous fat on the body is not clear, though it is known that the cells forming fat tissue show a vitamin complex. Perhaps it is a remaining stage from man's earlier history when it was necessary to carry one's own emergency rations. The female role of nursing mother would make this doubly necessary. The residual areas for body fat are: the buttocks and the flanks; the abdomen down to the pubic arch; the anterior and lateral surfaces of the upper thigh; the breasts or pectoral region; the back of the neck about the vertebrae prominens, and the posterior of the upper arm. (Illustration **13**.)

The pectoral region is the site of fat tissue which contains the mammary glands. The mammary glands commence development at puberty when the fatty tissue of the region also multiplies. The breast is at first disk-like, and next assumes a conical appearance and finally a hemispherical form. (Illustration **9**.) The breast tissue formation is without any structural support; because of this fact and the infinite possible range of breast proportions it is impossible to place much reliance on transverse circumference measurements taken at nipple section.

13
Resdidual areas for body fat.

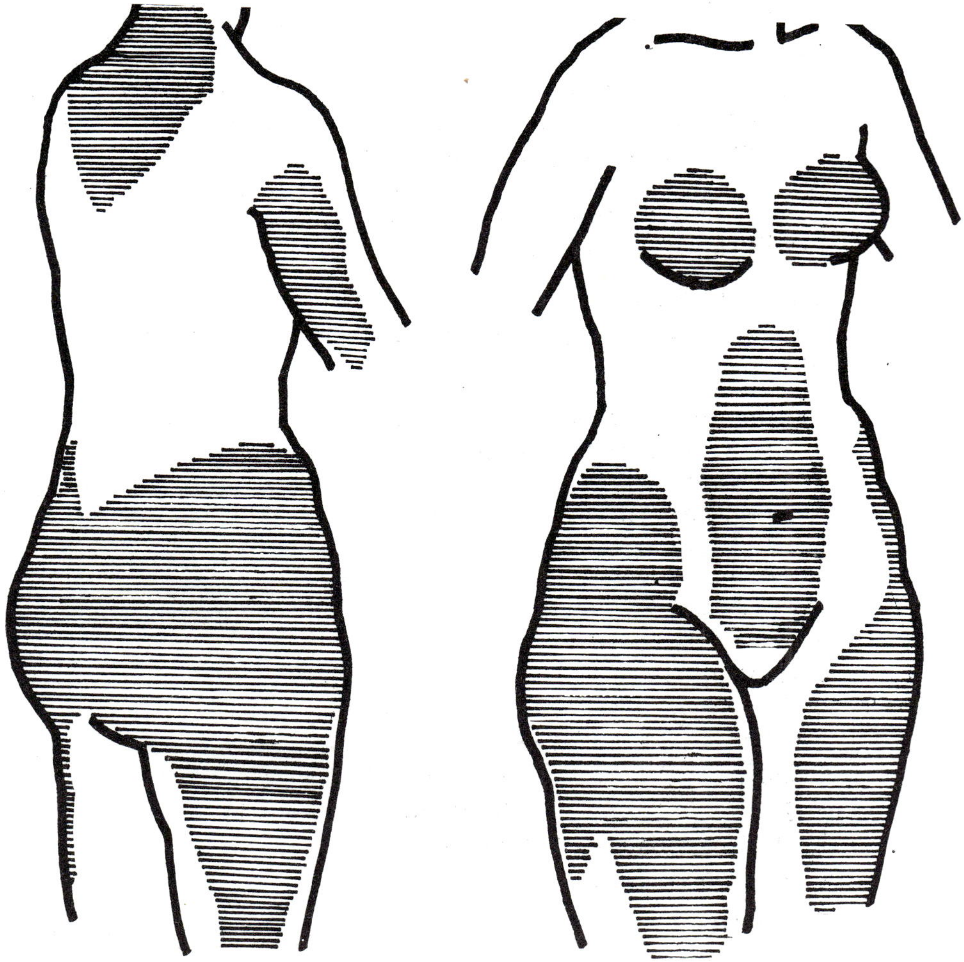

It can be seen from examining the list of residual areas of fat that the greater part of its accumulation is situated about the thighs and hips. Accumulations of fat are more readily noticeable in this region, and they are a typical female characteristic. The fat of this region has been found to have a higher correlation to body weight than the fat in any other region, on female adults. The heavier women will therefore always have the largest hips. Correlations of the fat of other body regions to weight are very much lower and have no reliability. In the more prosperous populations body fat increases in both sexes with age; the middle and early old age groups showing tendencies towards corpulence.

There is evidence of the effect of a superior geographical or social environment on total body growth. There have been indisputable increases in growth rates in the more prosperous countries, and there is a definite trend towards earlier maturation and greater total body build. This does not mean that a definite increase of nutrients will always produce a definite amount of body tissue.[1] Genetical and environmental factors combine to inject a series of unknowns into growth rates of children from a similar population; equally well nourished or not.

There are regional differences in growth that cannot be explained in terms of nutrition or climate, but would seem to depend on genetical factors peculiar to the region.[2] There can be some delay in growth rates of children in hot climates.[3] These differences may be in part responsible for the range of physique to be seen in man globally. The findings of anthropometric surveys in England of children at adolescence show that a superior environment can produce an increase in total height from one and a half inches (38 mm) to two inches (51 mm). An inferior environment might cause a loss

1 Tanner, J. M., *Human Biology*, Part IV, Factors affecting growth, p. 343.

2 Tanner, J. M., *Human Biology*, Part IV, Regional differences of growth, p. 345 to 347.

3 Weiner, J. S., *Human Biology*, Part V, Climate and growth, p. 457 to 459.

of expected height of one and a half inches (38 mm). During infancy loss or gain shown in total height, due to the same factors, is about a half an inch (13 mm). There is evidence that suggests these figures would be true for European and North American populations.

Because of hereditary and environmental influences we can find great variability in children's growth rates. Identical children, other than identical twins, do not exist; childhood presents a continuous variation of measurements. Average body measurements for children cannot be characterised by a mean value obtained against age. Numerical age has no great significance when evaluating children's sizes for design purposes. The two quantities needed are body dimensions and their velocities of growth (illustration **14a**.) A series of measures of one particular body dimension of a child taken over a period of time would show the velocity of growth of that dimension. A series of these measures obtained from a number of children, and then stated in graph form, and the curves matched, so that maximum and minimum velocities coincided irrespective of age, would allow an average curve or parameter to be drawn. (Illustration **14b**.)

The examination of average curves of children's body measures would enable a designer to determine where positive changes in terms of increases in dimension make a new size necessary to be fitted. He would as well have a record of the change of body dimensions. This method would be equally suitable in any sphere of children's requirements where a minimum or maximum range of design sizes needed to be calculated.

Ageing

The process of ageing is a gradual one and it is impossible to determine with any accuracy an average chronological progression of features which would be of much use. Old age is an easily recognisable phenomenon, but difficult to assess in terms of loss of potential or useful value.

Examination of bodily tissues shows that the ageing process is a gradual series of degenerative changes in the nature of morphological systems; like the processes of growth, the processes of degeneration in body parts do not necessarily proceed at similar speeds. Some morbid changes in structure precede others. One of the changes is the diminishing power of cartilage to maintain its elasticity. Calcification of cartilage becomes more pronounced with advancing age. Another is the increased fragility of the bony skeleton. The remodelling process whereby bone renews itself gradually ceases; bones become slighter and more brittle with age. The nervous system seems better able to maintain itself. Of all the different types of cell in the body, those of the nervous systems have the longest existence. So one can speculate that for a moderate period a comparatively stronger nervous system can

14a
This shows the manner of increase of arm length for five boys from five to sixteen years of age. The amount of total increase is similar through the range for all the boys, that is approximately eleven plus inches (279 mm). The difference of overall length before the growth noted here took place was four inches (102 mm). The smallest boy having an arm length of 15·5 inches (394 mm) approximately, the longest an arm length of 19·5 inches (495 mm) approximately.

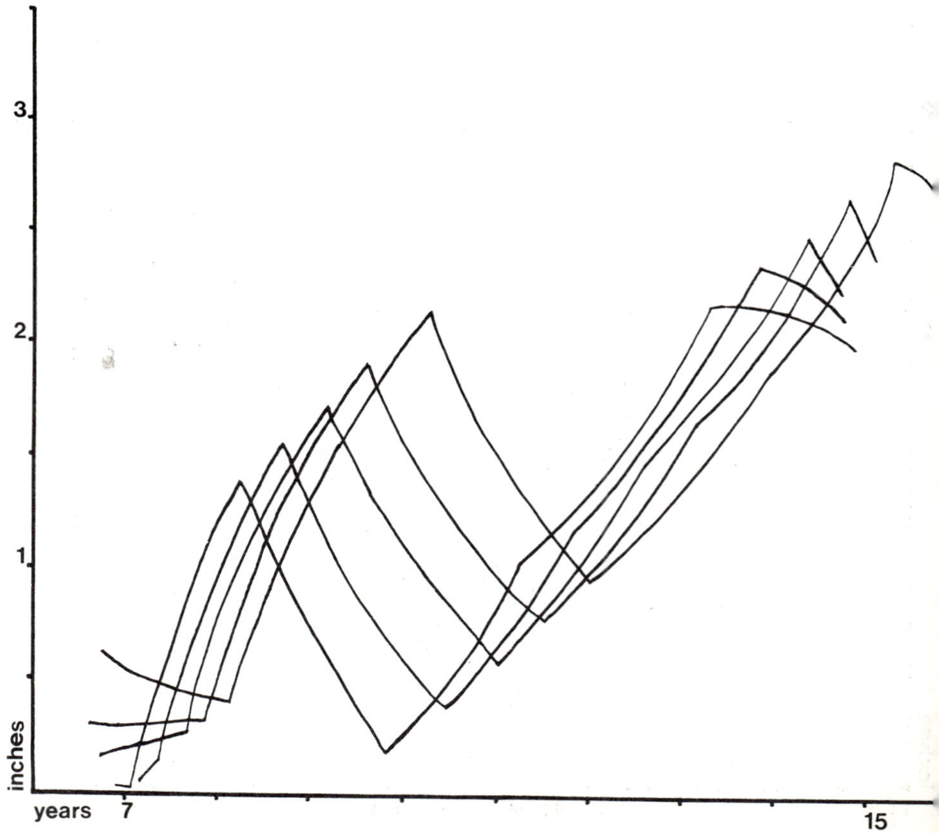

compensate to some extent for a less adequate mechanical system. However, after about 60 years of age there is a recognisable loss of motor functions, and a decreasing sensitivity to external stimuli.

Activity is gradually reduced and there is a period of natural decline of bodily functions; the period of natural decline now being 20 years or more.

Degenerative out-growths occur from the margins of bones, which together with other age changes in joint tissue, and the history of traumatic effects, makes movement more difficult. There is an increased incidence, with age, of a variety of forms of rheumatism, and other diseases of a crippling nature.

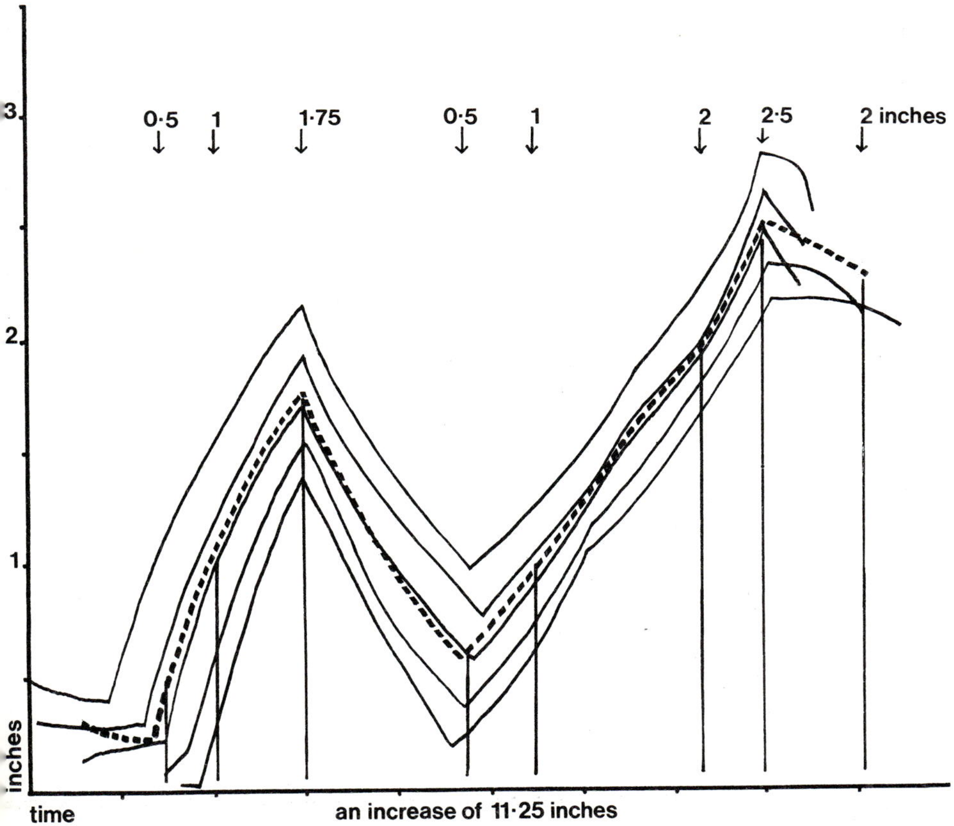

14b
The average curve here simplifies the estimation of the increase in the boys' arm length. The five curves have been moved so that their peaks and troughs more or less coincide. It can now be seen that the mode of increase for all the boys is similar although each one took place at a different age range.

The incremental additions chosen on this example are marked by perpendicular lines. The amount of the increment is given above the perpendicular line in each case. Other incremental points might have been chosen. Those chosen would depend on the experience of the designer concerned. For example a clothing designer would be able to select the number of sleeve sizes that were necessary to cover the mode of increase of arm length. Other dimensions of a growing child's figure could be studied in a similar manner.

The respiratory process and heart pump action also diminish in efficiency. All of these occurrences further restrict or slow bodily movement.

Outwardly old age shows itself in the face in a reversal to infantile proportions; mostly due to distortion by the over-growth and decay of the teeth, and malformations and wear in the jaws. The skin surface generally appears to wrinkle and sag and becomes coarser in texture on some parts of the body. On the face and backs of the hands the skin

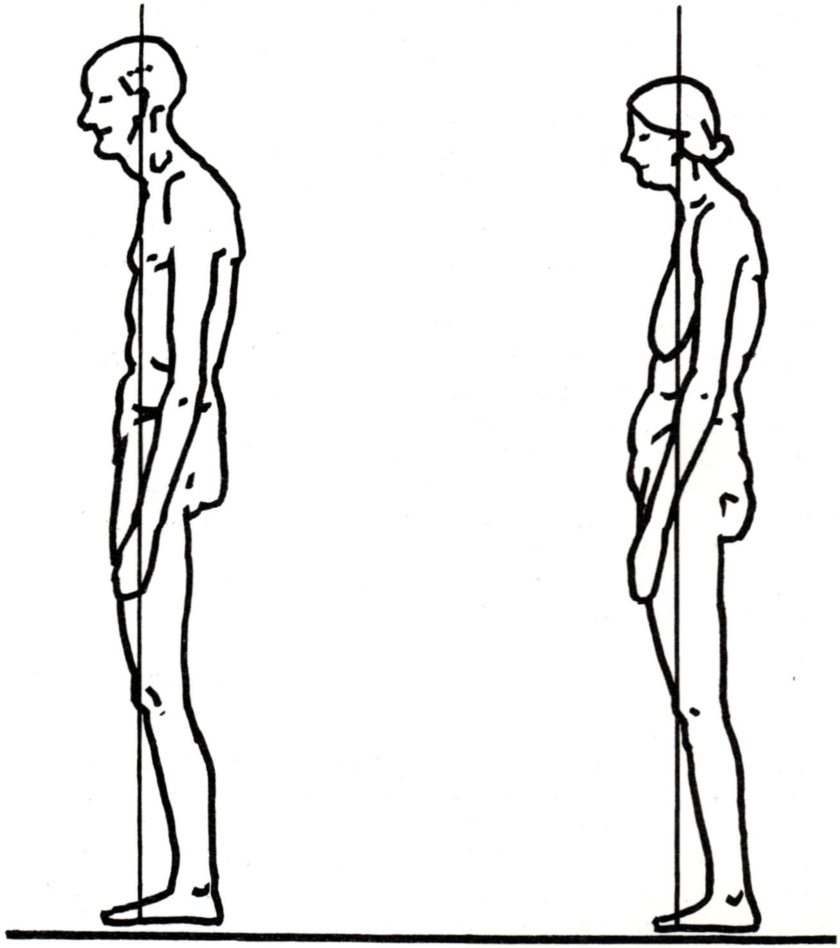

15
Posture of old age shown against the line of gravity.

decreases in thickness, but it increases in thickness over the thoracic spine, the heels and in the post-delford region. Fat may cause loose and unsightly folds. Veins are noticeably larger. Due to calcification or atrophy of intervertebral discs there is a loss of movement in the trunk as well as a loss of height, but this last point can be overstated. (Chart **7**.) There are changes in posture as a result of weakening musculature and joint distortions. (Illustration **15**.) The stooping look of advanced old age is accentuated by gross changes in the lower limbs and region of the hips. In the earlier stages of old age body weight often increases and this throws more strain on a weakening musculature and the legs, as well as on the heart functions.

Maintaining a working posture becomes increasingly difficult for elderly persons and they require a slightly higher work surface for comfort. They dislike bending.

Ageing has an insidious onset and in one sense it can be said to have commenced at any time after the peak of physical activity has finally passed in the late twenties and early thirties. Thereafter there is a very gradual decline of bodily powers, which is a part of a long chain of physiological events from birth to death. In general there is a 50% loss of strength between 30 and 70 years of age. Without the intervention of pathological causes the percentage loss of strength will not be very large until after 40 or 45 years of age. The percentage loss between 30 or 35 years of age in strength is small; and in terms of exact measurement it is extremely difficult to show the slight physical limitations for a particular person in this age range. In a normal healthy person, used to an active situation, it may still be extremely difficult to detect significant loss of physical powers up until the middle or late fifties.

Marked physical incapacities are easily recognised. But the slighter restrictions on human activity and their effect on performance are hidden from view by the human desire to exert maximum force and effort whatever the cost. We find with human performance that percentage rate of performance depends on the percentage rate of success achieved, depending again on the limits of activity particular nervous systems can tolerate.[1] Success often comes by a combination of a process of trial and error and a good memory. The older age groups can be rich in these qualities which we call experience and wisdom. Experience and wisdom find means of circumventing disabilities and restrictions, obscuring known physical limitations, and achieving success. Any fruitful investigation of declining bodily mechanisms and powers would have to be an association of the psychology of ageing with physiological and anatomical studies. Meanwhile we can say that the wisdom and intelligence gained through a working life can prolong its usefulness.

1 Thompson, d'Arcy, Ed. Bonner, J. T., *Growth and Form*, Camb. Univ. Press 1962.

Section 2 **Figure Typing**

All human populations show distributions of varying physiques. A variation of physique is represented in any age group and in either sex. Small differences in bodily proportions at birth are continuously multiplied by differential growth rates up until maturity. Man is evidently a type of animal which would appear to derive some advantage from propogating a range of body types. Part of physical anthropology study is concerned with the investigation of the variety of body types and anatomical peculiarities, physical evolution, physical growth, maturity and ageing and their characteristic patterns. Some of the characteristics are quantitative like those of length or girth or weight. Some are qualitative like those of type of body hair or skin colour. It is the field of quantitative characteristics that is the special concern of anthropometry for designers or applied physical anthropology. Quantitative characteristics include all external measurements of human body dimensions. Other quantitative characteristics that can be measured include angular and linear ranges of body segments and postural and limb forces and strengths. These two sets of quantitative characteristics of body measurements are sometimes called static and dynamic anthropometry.

When we examine a range of human physiques we are left with an impression of varying degrees of muscularity, of fat or corpulence, of small wrists or long thin feet, and so on. It can be seen that measurement of body parts is a formidable task. During this century and the last surveys and anthropometric field work have collected a great deal of metrical information about differences of body physique and constitution. Although precise measurement of physique is relatively modern, since the beginning of Greek civilisation there has been a curiosity and a cultivated interest on man's part about his own shape. The progress of this understanding has intermingled with man's quest to understand his own substance and its history.

One of the most important aspects of this understanding has been the development of human biology as a separate discipline. In the development of diagnostic practice in medicine from the time of Hippocrates a number of attempts have been made to classify the inter-relationship of elements that determine a body type. An important part of the examination of body constitution has always shown concern for enquiry about body shape. A number of attempts have been made to classify body shapes into types which share common normal or pathological features. Outside the medical field these classifications have been of continued interest and benefit to anthropologists and designers. Leonardo da Vinci attempted to correlate body segment proportions by geometric methods, showed interest in biometry and assisted the advancement of anatomical studies in his day.

Hippocrates determined by observation two main body types: the *phthisic habitus* or the thin and tall physique, and the *apoplectic habitus* or the thick and short physique. Very notable developments in this field of body-typing have been made by an Italian physician, Viola, and a French physician, Rostan, and by a German psychiatrist, Kretschmer.

Thirty years ago William Sheldon, an American psychologist, after much patient research and brilliant analytical thinking, devised what might well become the classical method of body-typing. Although doubts have been cast on some aspects of Sheldon's method the feasibility and usefulness of it has been proved, in particular by Dr. J. M. Tanner who has refined and advanced the use of the system.*

After Hippocrates subsequent methods of figure-typing still adhered to the formula of a number of rigid divisions, and to the idea that all human physiques could be divided simply and placed into a few categories.

Hippocrates' division was expanded into three, four or five main body types. Only Viola acknowledged the presence of a 'mixed-type' of physique that contained some of the elements of the three primary groupings he had devised.

* Tanner, J. M., *Physique of the Olympic Athlete* (see Bibliography).

His three primary groupings also indicate his objective concern for the mixture of component features seen in any single physique. He saw a Longitype with long limbs and a wide thorax, and a Brachitype with exactly opposite proportions: the third type, the Normotype contained all those physiques shading between the first two types.

The fourth, and most significant, mixed-type contained all the physiques whose anomalies excluded them from a place in the first three groupings. The mixed-type is rather more the rule than the exception in the study of physique. With the conception of the 'mixed-type' there was the idea, current for the first time, that there can be anomalies occurring in some of the components of a physique which do not place them in an abnormal category.

Sheldon's system took the concept of the mixed type of physique further in that he stated that any physique contained three components, but in a varying ratio. These variables are continuous in any population.

William Sheldon's research was carried out at Harvard University prior to 1940, when his findings were published. His specimen population consisted of 4,000 students. He used a combined method of photography and anthropometry. Each subject was photographed inside, front and back view. Inspection of photographs was used in conjunction with selected anthropometric measures and indices.

For this kind of photographic work the subject has to be posed in a careful manner, and the working distance between the camera and the subject has to be of a special length to avoid perspective distortion at the perimeter of the image. The camera type itself must be carefully chosen. Sheldon observed that there always appeared to be three aspects determining the morphological structure of an individual, and that they were apparent in any body region. These three aspects or components of bodily morphology he called endomorphy, mesomorphy and ectomorphy. The three descriptions were devised from the terms used to describe the three initial layers in the early embryonic forms of higher life.

Endomorphy, as a component, describes a predominance superficially in fat giving a roundness and fullness of shape. In deep bodily morphology it indicates a large digestive system. With a round head and weak appendages the endomorph retains something of an infantile appearance. Mesomorphy, as a component, describes a predominance in muscle and bone, giving an angular and hard shape. In the bodily morphology generally it indicates a strong bony frame and connective tissue. With little subcutaneous fat and a resolute stance, the mesomorph is very much the ideal adult and the person of action. Ectomorphy, as a component, describes a predominance of skin surface relative to body mass; ectomorphy in extreme shows long thin body segments giving a weak and poorly balanced posture. In deep bodily morphology it indicates a large central nervous system and brain. The thin appendicular skeleton by contrast makes the trunk look too small and the head too large, and there is a consequent suggestion of immaturity in this body shape. A list of the predominant differences between endomorphy, mesomorphy and ectomorphy are listed in Chart **9**.

The origin and connection between Sheldon's components and embryonic layers may be tenuous and beyond proof, yet they give the best means available for evaluating body shape. Any whole physique, or any anatomical segment of a physique, can be represented in terms of a combination of three components. A short training would enable a designer to think in terms of these components and develop an awareness of the nature of the diversity of human physique.

Sheldon assigned a scale of numerals for each of his three components. By anthropometric method[1] and visual photographic inspection he established a scale of numerals from one to seven in each component. The scale of numerals was equally spaced and became ratings in the system which he called somatotyping. The scale of numerals shows the strength of any one of the three components in any physique.

1 Each fresh somatotype when rated was compared with other somatotypes whose subjects shared a similar height to cube root of weight ratio.

Somatotype ratings give the component of endomorphy first, then that of mesomorphy and finally ectomorphy. So a somatotype rating of extreme endomorphy would appear as 7–1–1; a somatotype rating of extreme mesomorphy 1–7–1; and a somatotype rating of extreme ectomorphy 1–1–7. These extreme somatotypes are rare. There are only approximately five present in any population unit of a thousand persons. More common somatotypes present in any population are 3–3–4 or the 5–3–3. These last specimen ratings show how the components express the continuous variable nature of physique, and sheds some light on the interdependence of the components as factors. Chart **10** gives the approximate incidence of a variety of somatotypes in a population.

CHART 9 *Possible dominant components in somatotyping system.*

Endomorphy. (Male and female illustration **16** and **17**.) Total physique. Round and soft with large fat storage. In extreme form pear shaped. Abdomen is full and extensive and the thorax appears small. The limbs appear short and ineffective. Shoulders are full and round supporting a rounded head. The skin is soft and smooth. Fine hair, with little showing on the body.

Endomorphic Skeleton
In side view the vertebral column appears straightened. Bony girdles of trunk and pelvis are approaching the circular. All the bones are small with tuberosities and projections rather rounded.

Head and Trunk
Features small and unobtrusive. Small cranium in relation to wide palate and face and a short neck. The trunk is long and heavy at the base. The greatest body breadth is near the waist which is difficult to determine. The thorax is wide at the base. The whole trunk and head has a forward pushing penguin-like appearance.

Arms and Legs
The limbs are short and from an inflated appearance proximally they rapidly taper down to small and weak distal extremities. There is no angularity, and so in the male where there can be the appearance of breasts, there is a suggestion of femininity.

16
Extreme endomorphic components.

Mesomorphy (Male and female illustration **18** and **19**.)
Total physique. Square and rigorous in appearance with much prominent muscle. Shoulders predominate with thorax wide at

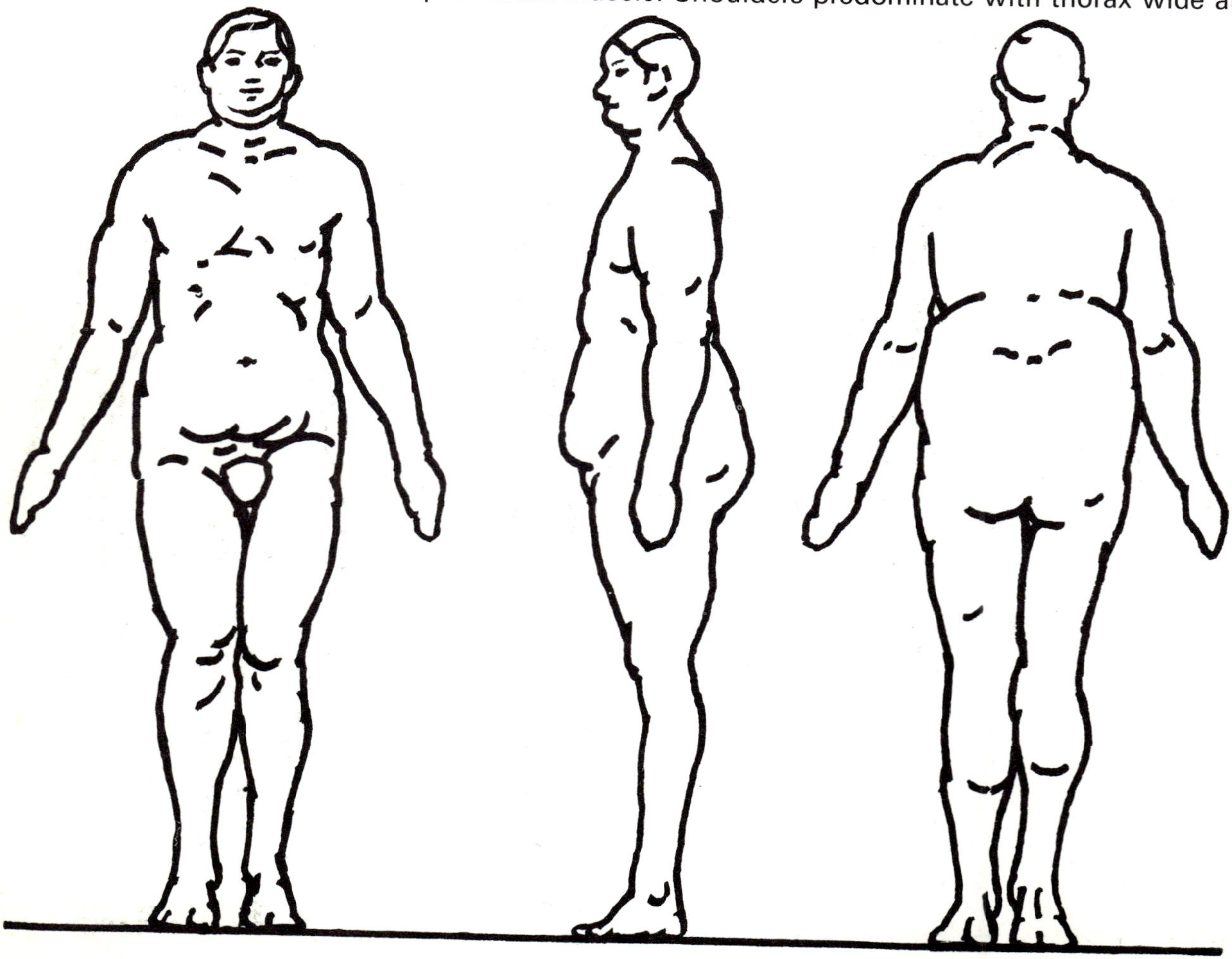

the apex, and the abdomen small. The limbs appear large and strong. Neck is strong with a rugged head showing prominent eminences. Skin is rather coarse, as is the hair.

Mesomorphic Skeleton
All the curves of the vertebral column are well shown. Bony girdles of trunk and pelvis are wider laterally than in the anterio-posterior dimensions. All the bones are heavy, with all the tuberosities, ridges and processes well defined.

17
Extreme endomorphic components.

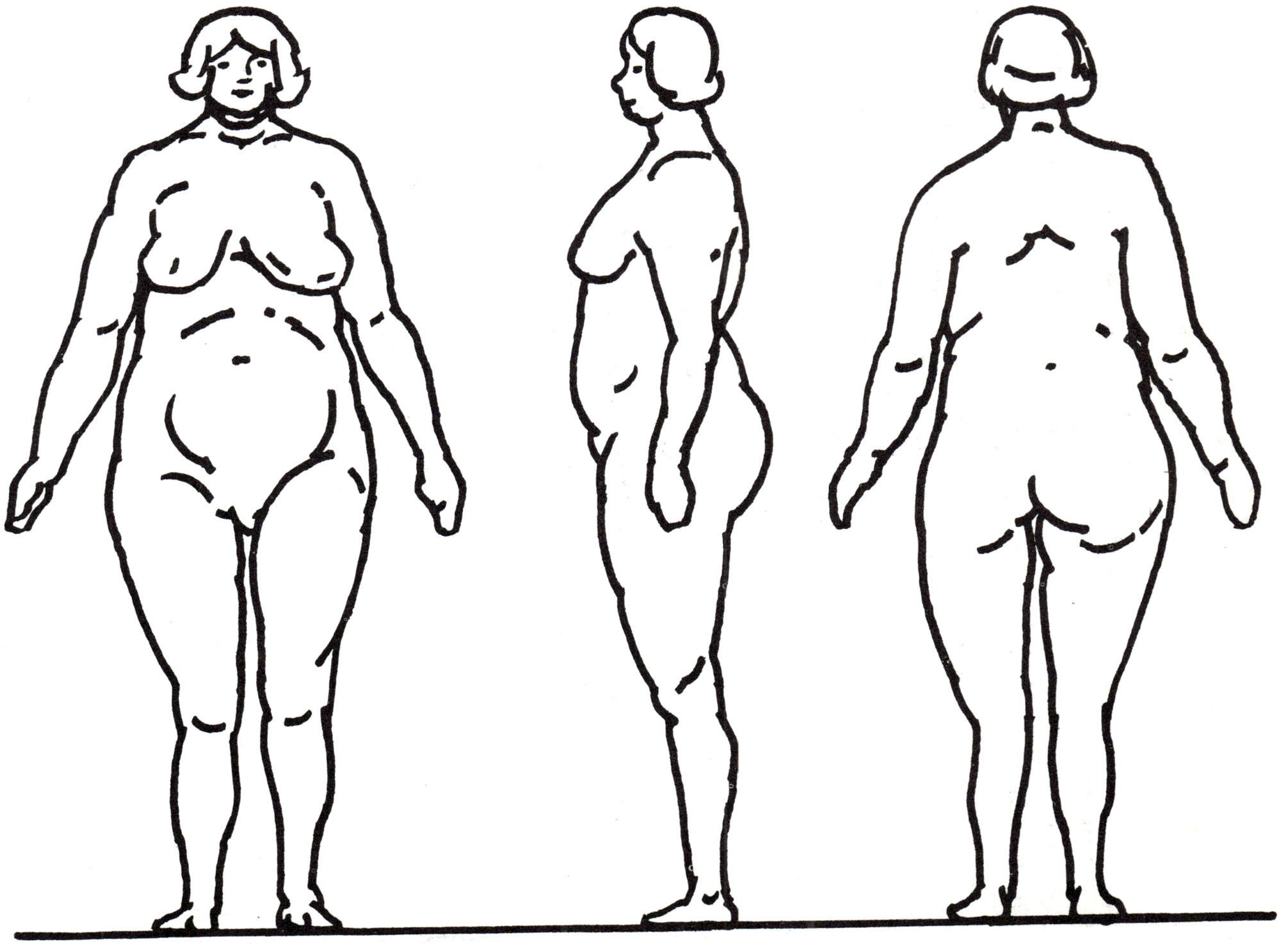

Head and Trunk
The face is large compared with the cranium. The forehead is often shallow and the facial eminences and angularities tend to approach 90°. The neck is long and the sterno-mastoid triangle well marked. The thorax is massive, and certainly in a man, much broader than the pelvis in the region of the xiphoid process. The waist is low with the abdomen almost perpendicular in profile.

18
Extreme mesomorphic components.

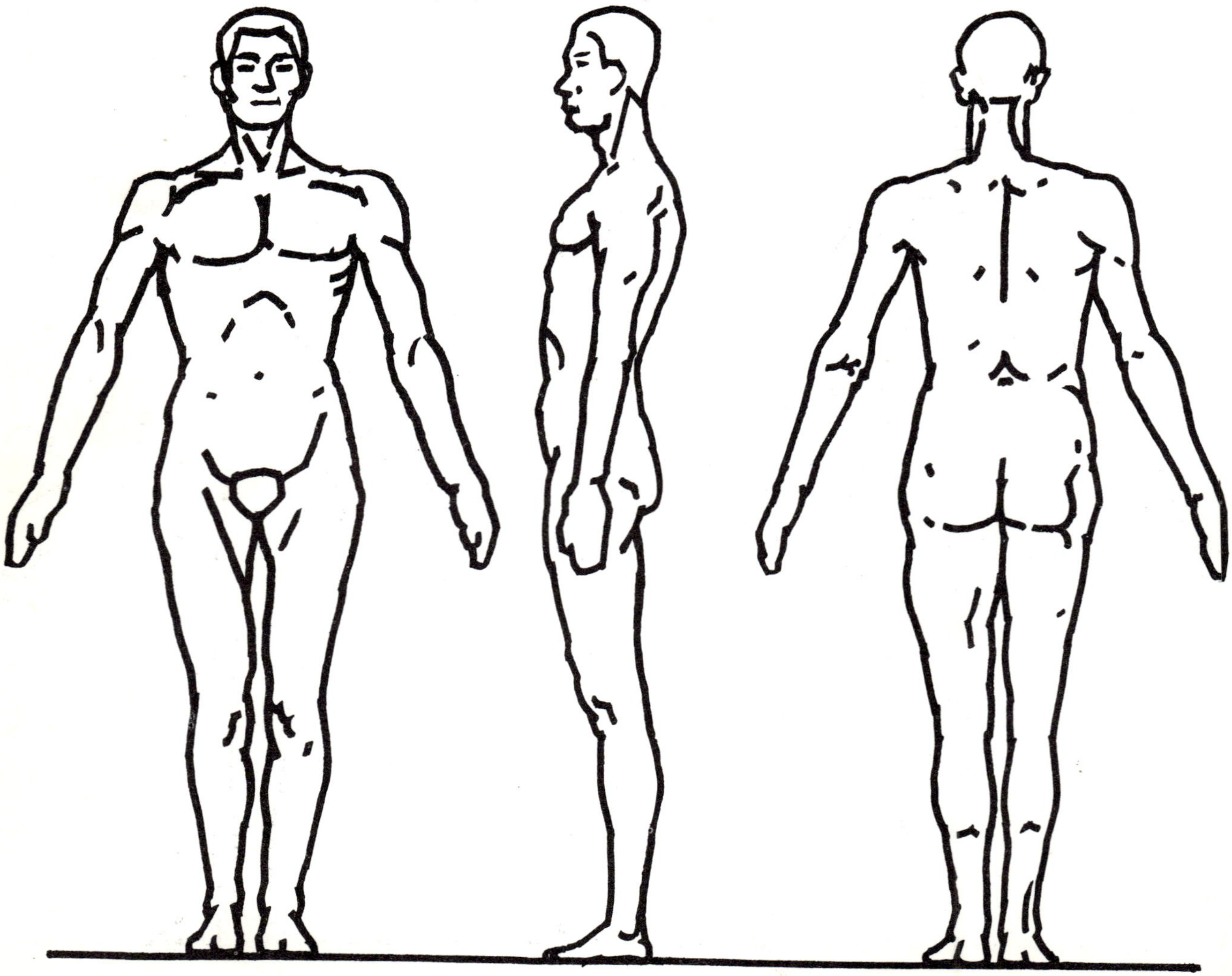

Arms and Legs
These are generally heavy and well muscled, particularly at their proximal ends. The hands and feet are also large and capable in appearance. Often the forearm and calf muscles are very prominent.

Ectomorphy (Male and female illustration **20** and **21**.)
Total physique. Fragile and slender with a minimum of either

19
Extreme mesomorphic components.

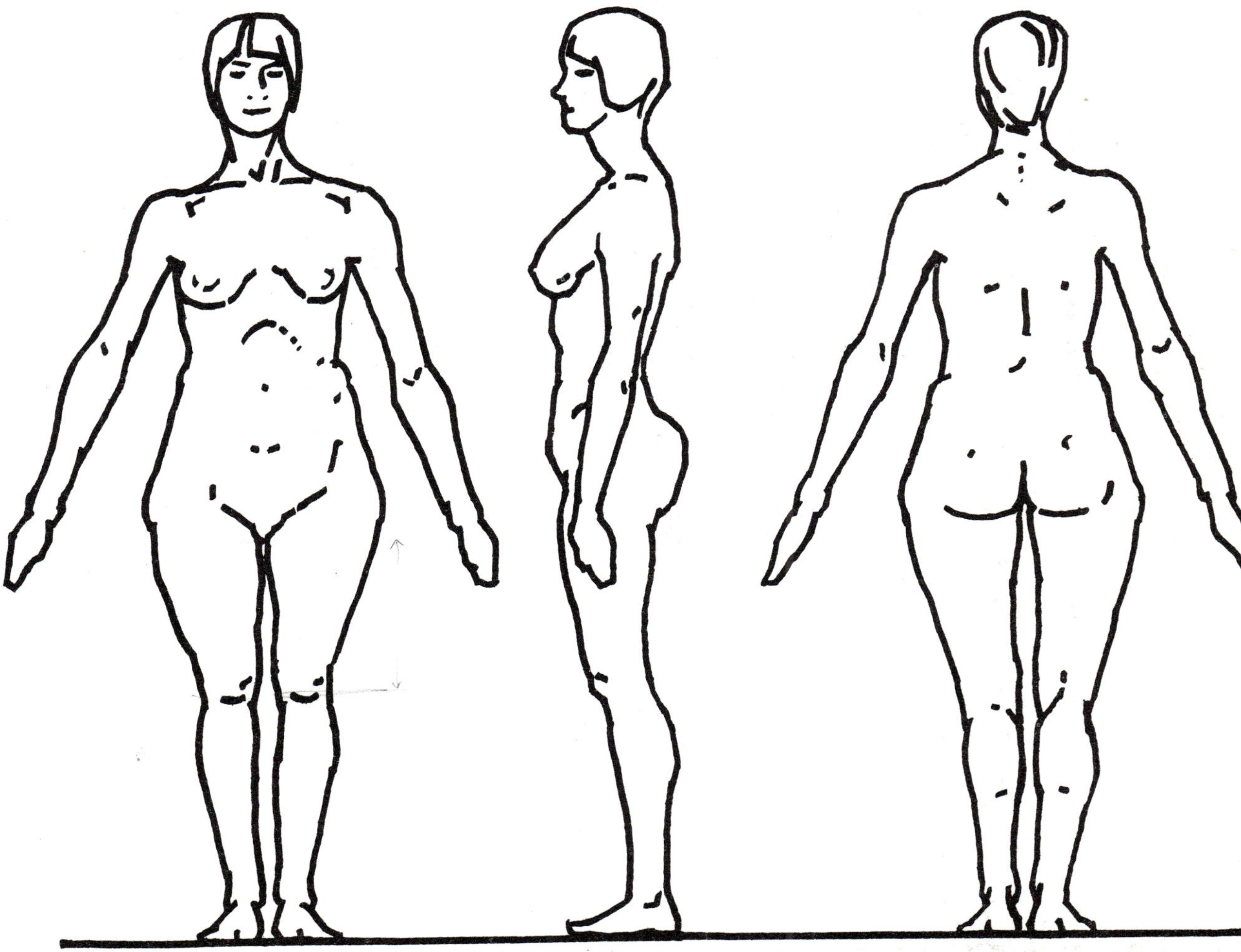

fat or muscle. The trunk generally appears short and poorly positioned and is accompanied by long spindly limbs. Shoulders are wide but droop, the neck is slender and can appear inadequate for the head, which often has a large cranium. The skin is thin, hair is brittle and very variable in amount.

20
Extreme ectomorphic components.

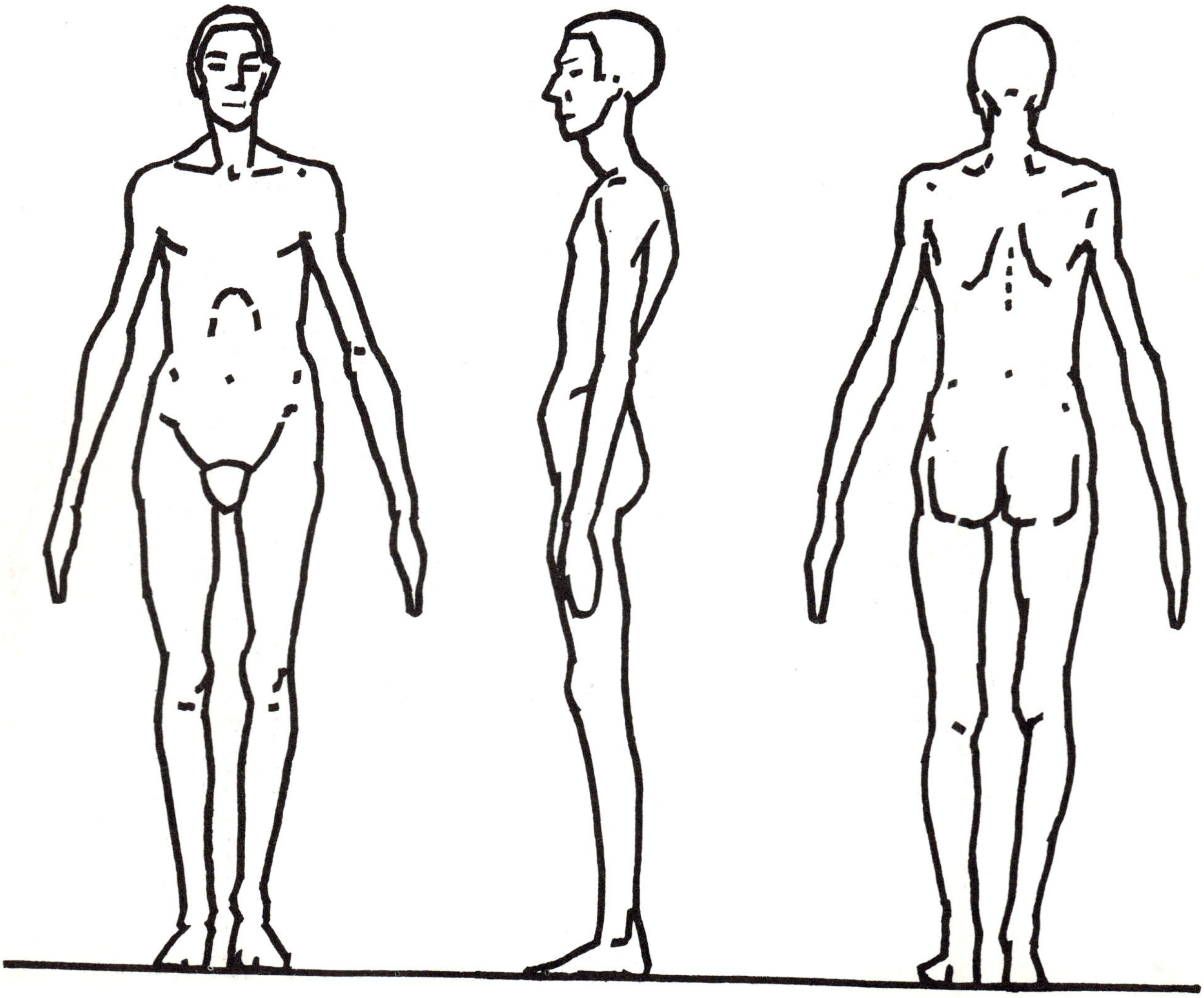

Ectomorphic Skeleton
In the vertebral column the cervical and thoracic curves are well marked to the point of distortion, whilst the lumbar curve is flattened. The sub-costal angle is acute although the thorax is flat, particularly in the region of the sternum bone. Bones are light, but also variable in length for stature for this type has a wide range.

21
Extreme ectomorphic components.

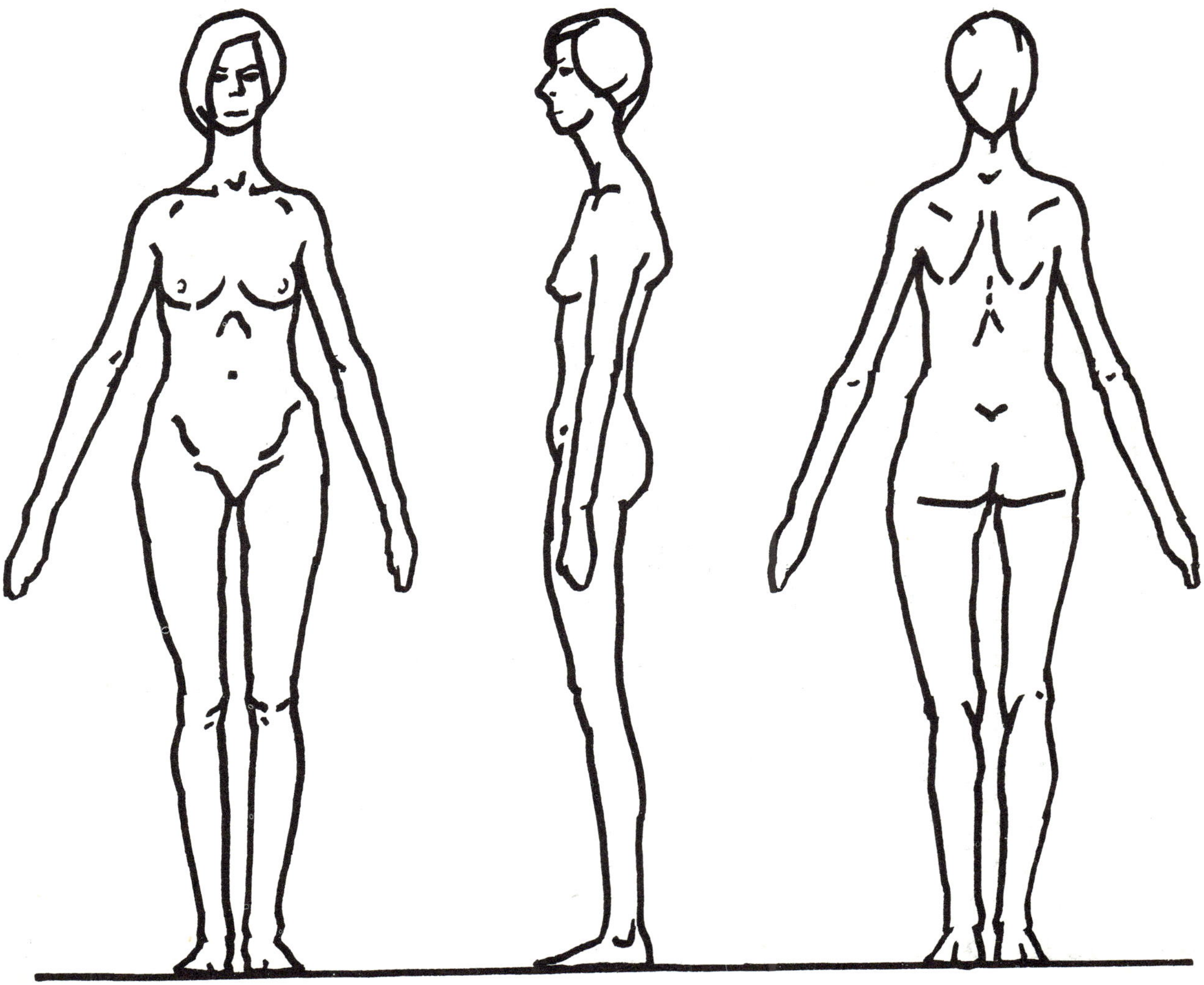

Head and Trunk
Features are finely etched with a slight chin. Brow can be large, the cranium and brain very extensive. The thyroid gland is prominent, and the neck hangs forward from rounded shoulders. The trunk is shallow and flat, the abdomen protruding in front of the thorax in profile. The distinct sag forward of the abdomen is due to ineffective musculature.

Arms and Legs
They appear weak and long with a slight musculature. Their weakness is more noticeable at the proximal ends; and distally the toes and fingers are long and delicate.

These descriptions are a caricature of the components. The majority of body-builds will show random samples of the components in a subdued form. Somatotype rating 1-7-1 is not known. The nearest to it, rating 2-7-1, is extremely rare. Ratings 7-1-1 and 1-1-7 are also extremely rare. Popular subdued forms show a middle range of mesomorphy with an emphasis towards a slightly higher range of ectomorphic features. Extreme endomorphy is not popular; in modified form it still has an occurrence of less than two per cent, and will appear in the guise of ratings of 5-3-2, and 5-3-4.

Some observations on chart No. 10 and figure-type facts.
Looking over all the columns it can be seen that low ratings are more common than high ones. A high rating of either six or seven will only appear in one component. Of the more popular types in columns (a) and (b) none add up to more than eleven. In columns (c), (d), (e), (f) and (g) none add up to more than 12, but some add up to only nine. Columns (a) and (b) therefore represent the more balanced figure types. Column (a) shows the more common male figure type, and the bottom of column (b) shows the more common female figure type.

Amongst the rarer types the bottom of columns (e) and (f) and the top of column (g) represent physiques more likely to

CHART 10 *Probable representation of figure types in a population of four hundred people*

(a) over 20 of each	(b) No less than 8 no more than 20 of each	(c) No less than 5 no more than 8 of each	(d) No less than 3 no more than 5 of each	(e) 2 or 3 of each	(f) 1 or 2 of each	(g) 1 of each
3.3.4.	2.2.6.	2.2.5.	2.5.2.	1.2.6.	1.1.7.	1.3.6.
3.4.4.	2.3.5.	2.4.5.	3.4.5.	1.2.7.	1.6.3.	2.1.7.
4.4.3.	2.4.4.	2.5.4.	3.6.2.	1.4.5.	2.1.6.	3.1.6.
	2.5.3.	3.5.4.	4.2.5.	1.5.4.	2.2.7.	3.7.1.
	2.6.2.	4.2.4.	4.5.3.	1.6.2.	2.7.1.	3.7.2.
	3.2.5.	5.3.2.	5.3.4.	1.7.1.	2.7.2.	4.1.5.
	3.3.5.	5.4.2.	5.4.3.	1.7.2.	3.2.6.	4.6.1.
	3.4.3.			2.3.6.	4.5.1.	4.6.2.
	3.5.2.			2.6.1.	5.2.2.	4.7.1.
	3.5.3.			2.6.3.	5.4.1.	5.1.4.
	4.3.4.			3.6.1.	6.2.1.	5.1.5.
	4.4.2.			4.3.5.	6.2.3.	5.5.1.
	4.4.4.			5.2.3.	6.3.1.	5.5.2.
	4.3.3.			5.2.4.	7.1.2.	5.6.1.
	4.5.2.			6.2.2.		6.1.2.
	5.3.3.			6.3.2.		6.1.3.
						6.4.1.
						6.4.2.
						7.1.1.
						7.2.1.
						7.2.2.
						7.3.1.
						7.3.2.
						7.4.1.

be masculine; the bottom of column (g) represents physiques more likely to be feminine. Nevertheless, none of the figure types represented are exclusive to either the male or the female.

Athletes are often found amongst the types in the top of columns (b), (c) and (d). 2.5.2. at the top of column (d) being particularly notable in this field and showing an outstanding harmonious physique.

Finally, amongst the rarer physiques there is perhaps a drift towards endomorphy.

22

3.5.3. This is shown to represent a somatotype common amongst men. This is a strong and muscular physique but not 'chunky'. The muscles show in slight relief. The head is firm with a well modelled face. The neck and shoulders are prominent and well displayed. The trunk exhibits an ample thorax and a neat abdomen and pelvis.

In recent usage the whole units of the original scheme have been expanded with half units, and 13 degrees are used for each component rather than seven. A somatotype might thus appear as $2\frac{1}{2}$–$4\frac{1}{2}$–4 or $1\frac{1}{2}$–4–$4\frac{1}{2}$. As a system somatotyping needs a good deal of practice; but there is very high level of agreement on ratings between practised users. Detailed knowledge of anatomy is necessary because ratings give scores for skeletal framework, bone widths and lengths, muscle size and fat thickness, as well as comments on the

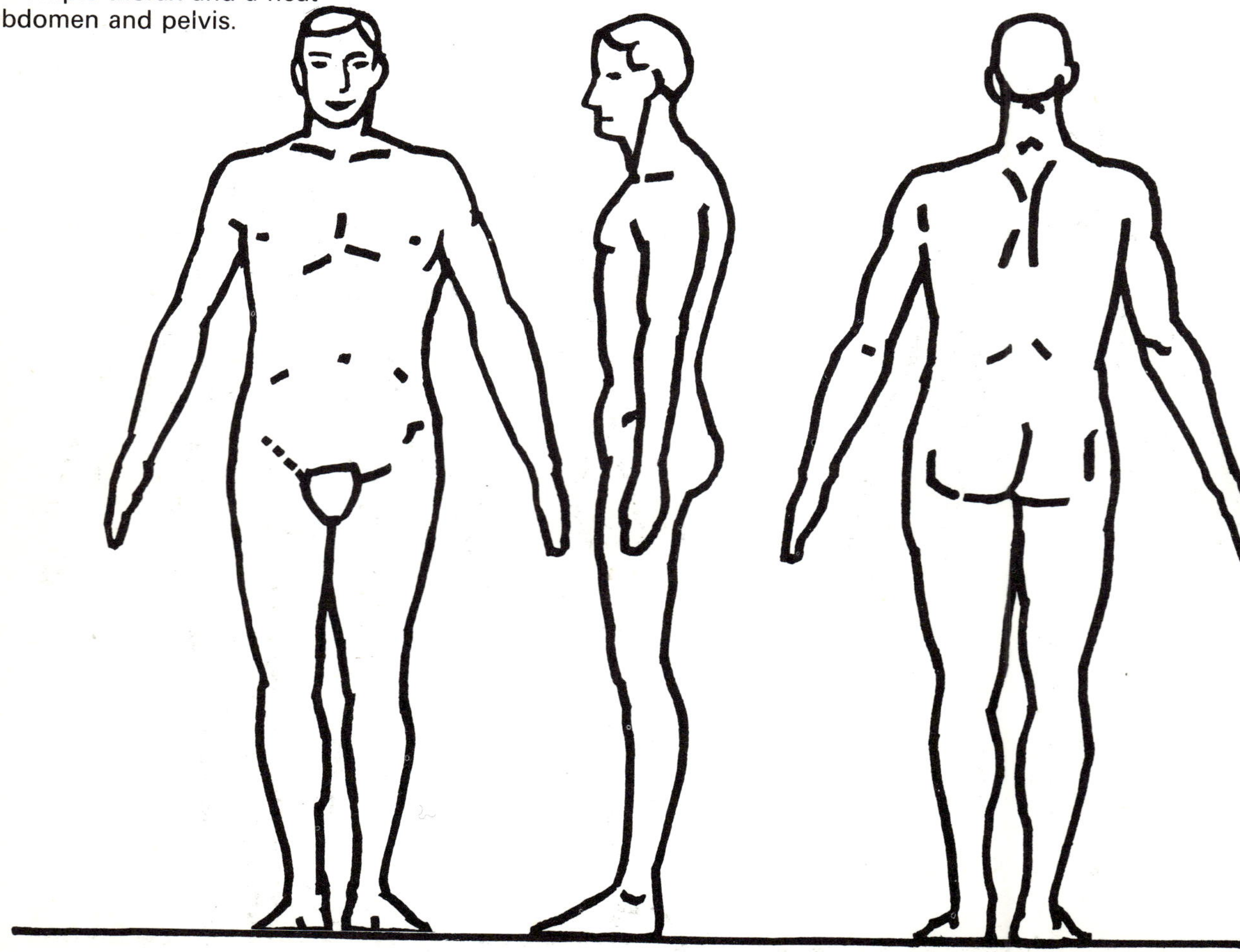

more superficial aspects of body-build. See lists of dominance). The somatotyping system can be used successfully for any cross-section of a population from any geographical region. It can also be used equally successfully with either sex.

Differences that do occur between the sexes are accounted for by degree ratings. Sheldon found that although man's ratings are between one and seven, women more rarely reach seven in any component. The intervals between ratings

23

5.3.3. This is shown to represent a somatotype common amongst women. The head is average to large; the shoulders small, and the waist well defined in the centre of the trunk. Endomorphic softening is present throughout, though the three in ectomorphy represents the longish limbs which are delicate at the extremities.

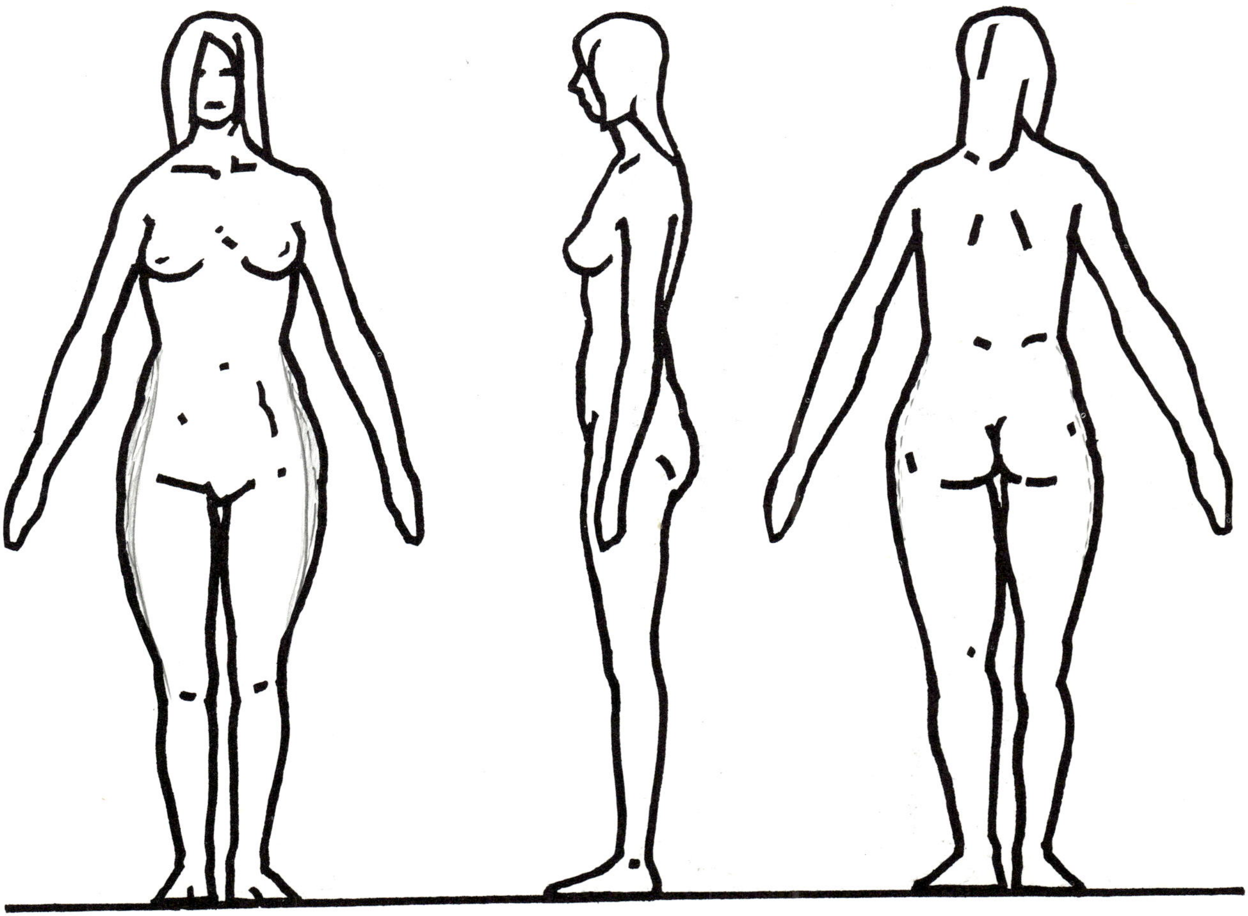

remain the same for both sexes; the ones, twos, threes, fours, fives and sixes of a component being similar for a man or woman. He found 3–4–4 or 3–5–3 were common ratings for a man, and 5–3–3 common for a woman. In men 5–2–3 or 6–3–3 was rare, and in women 3–4–4 was rare. (Illustrations **22** and **23**.) The physique of a mature person between 20 to 30 years has the best construction for

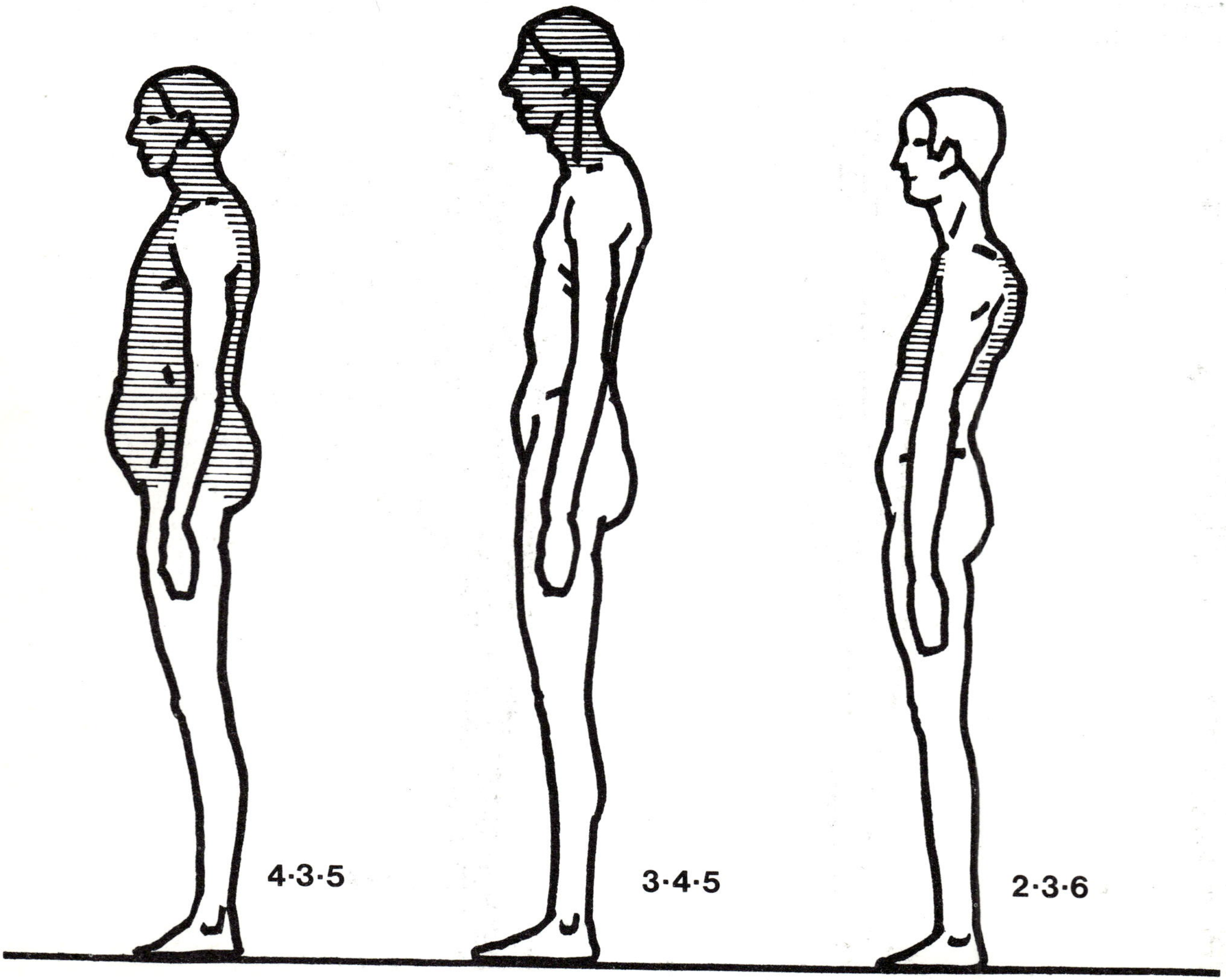

somatotype rating. Somatotype ratings are supposed to hold for life, and even if it can be proved that they do not the system is not invalidated from a designer's point of view. They are progressively more difficult to determine during obvious ageing and when pronounced pathological indications are present. For example, a type of obesity occurring in a predominant ectomorph might produce so much fat as to nearly obscure the true picture. But examination of the skull, side and front views of the thorax, the wrists and hands, and the ankles and feet should reveal the prevailing ectomorph.

24 and 25
To show examples of possible dysplasies in men and women's physiques.

4.3.4. Endomorphic legs together with ectomorphic head, trunk and arms.

3.2.6. Too small a head for the other parts of the figure which is long and lean.

4.4.4. A heavy mesomorphic head, with a mixed endomorphic mesomorphic trunk supported on ectomorphic legs.

4.3.5. An endomorphic head and trunk with ectomorphic limbs.

3.4.5. A mesomorphic head and neck supported by an ectomorphic frame beneath.

2.3.6. A small physique infected with a collapsed chest.

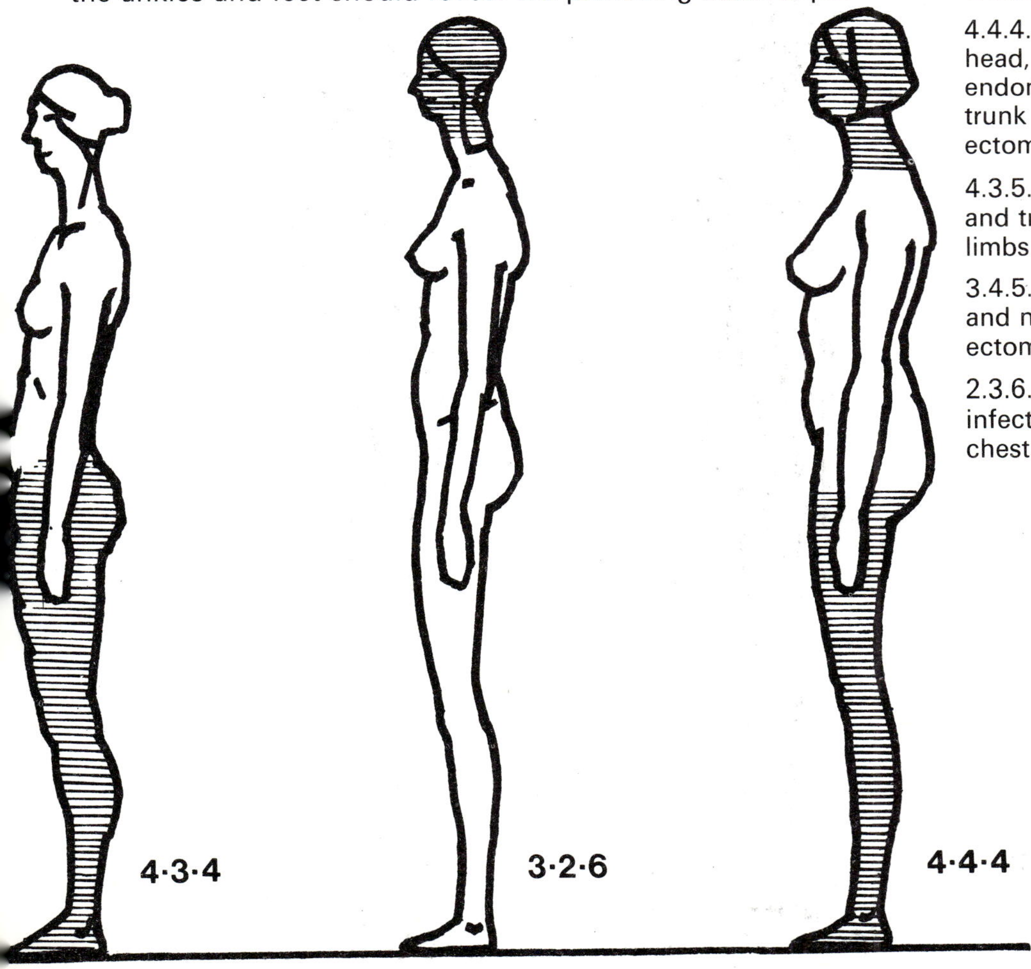

The somatotype scores illustrated here have been built up from a number of physical features originally distinguished by Sheldon, and given further definition by Sheldon, Tanner and others.

They have been put together by the identity kit method. For this reason the total figure silhouette and posture may not look entirely normal, although it is hoped that they caricature the figure type.

The presentation of these physical features in this form, for both sexes, has been arrived at by observations made during the direction of life drawing and anatomical studies. Any shortcomings are, therefore, entirely my own. (Illustration **26**, **27** and **28**.)

Tanner has said that it might be possible to evaluate quite a young infant's somatotype when enough data has been collected. The system can be used with some success from a 12 year old onwards. When somatotyping growing children it has been found that the mesomorphic component is often dominant; more particularly so in middle and late adolescence.

Body types emerge during the pre-puberal as well as the adolescent period of growth. There may be slight changes in a developing child's somatotype because of hereditary factors. Therefore, phenotype becomes a more useful term when discussing children's bodily constitution. (Phen derives from the Greek verb *phainō*, meaning to appear. We thus have the description for children; the type that is making its appearance.)

Sheldon observed, like Viola, proportions in some adult bodily regions that were not compatible with proportions somatotyped in other regions. They formed disharmonies in physiques, and were lawless in relation to the rating which the majority of a physique might warrant. This appearance of disharmony in a physique he called a dysplasia, defining them as that which were out of order. (Illustrations **24** and **25**.)

A high endomorphic rating for thighs and legs with an ectomorphic trunk and arms is a typical dysplasia often seen

26

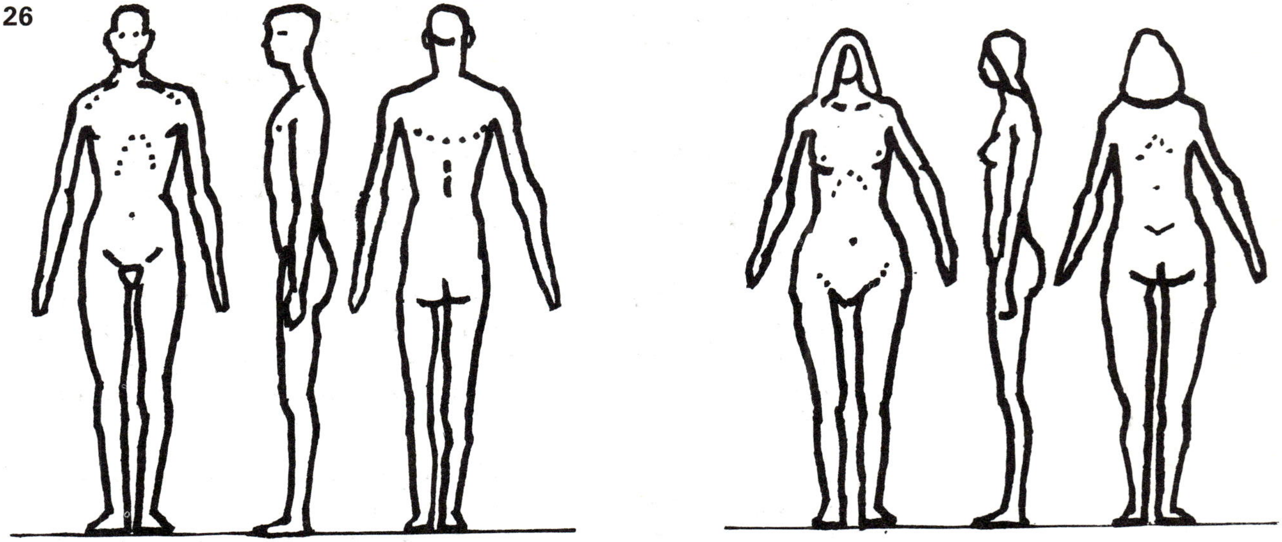

rating 2·3·5

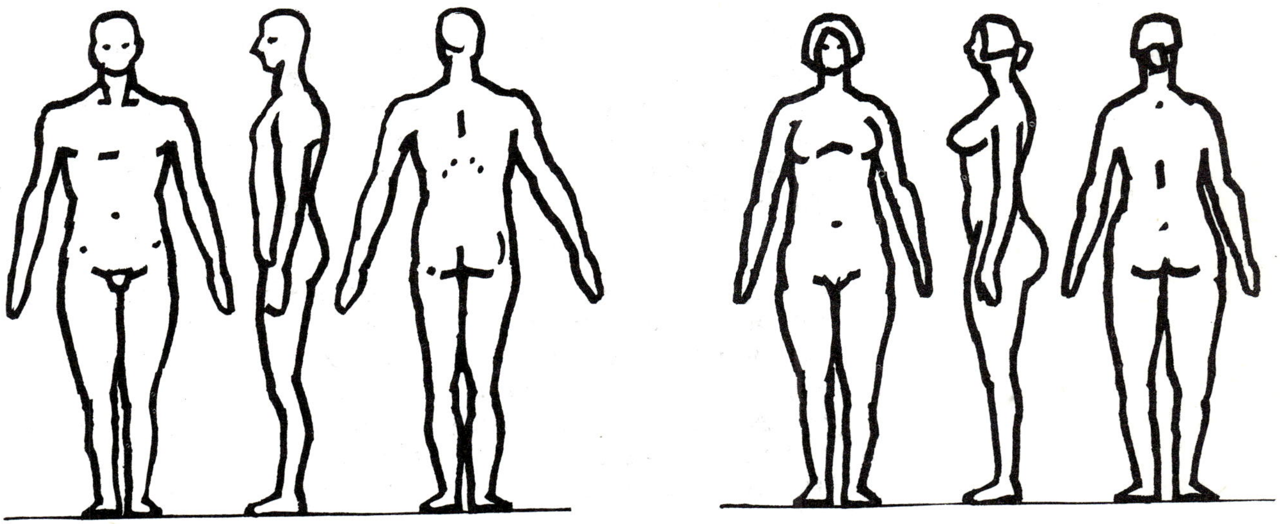

rating 3·6·2

in a woman. A mesomorphic head with a trunk and limbs ectomorphic figure is a dysplasia sometimes seen in men.

Pathological conditions may also present other forms of dysplasia. There may also be the presentation of bisexuality in physique. Gynandromorphic scales were created by Sheldon for use in fields of study where a system of measurement was needed for sexing body segments.

Illustrations **26**, **27** and **28** are cameos of a few types of physiques with their somatotypes, given by the Sheldon method.

1.4.5 This shows an interesting combination of mesomorphy and ectomorphy. An unusual and impressive physique.

2.3.5. This is a well proportioned physique although the skeleton is light. It is fairly common and although suggestive of power it is not an athletic figure type.

4.1.5. This is a quiet physique with little muscular control. The skeleton is weak and might either be long or short. The female of this type shows the more obvious proportions of her sex, the lack of mesomorphy suiting her.

6.3.2. Is one of the giants among figure types. There is a marked endomorphic pneumatic appearance, particularly in the abdomen and the thighs. Although the limbs are well covered with fat in their proximal portions, the limbs are quite heavily muscled. It is a tall type with obvious neck, shoulders and trunk.

3.6.2. Powerful and chunky figure type. Prominent bones and muscles. A physique to be found at athletic meetings.

5.4.2. One of the nicer endomorphs and not uncommon. Not a tall type to which an 'inflated' trunk gives a slightly 'dumpy' appearance.

From the design standpoint it is safe to consider that any bisexuality can be explained in relation to a specific somatotype. There are some regional differences in physique. It is important to realise that in the course of a survey of a population of any particular region, greater differences are

27

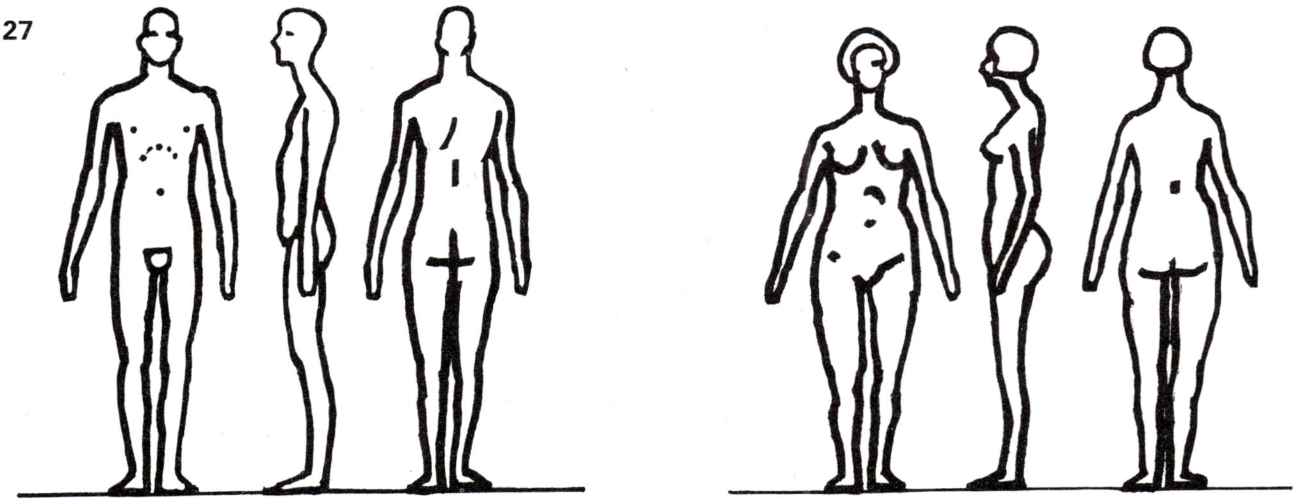

rating 1·4·5

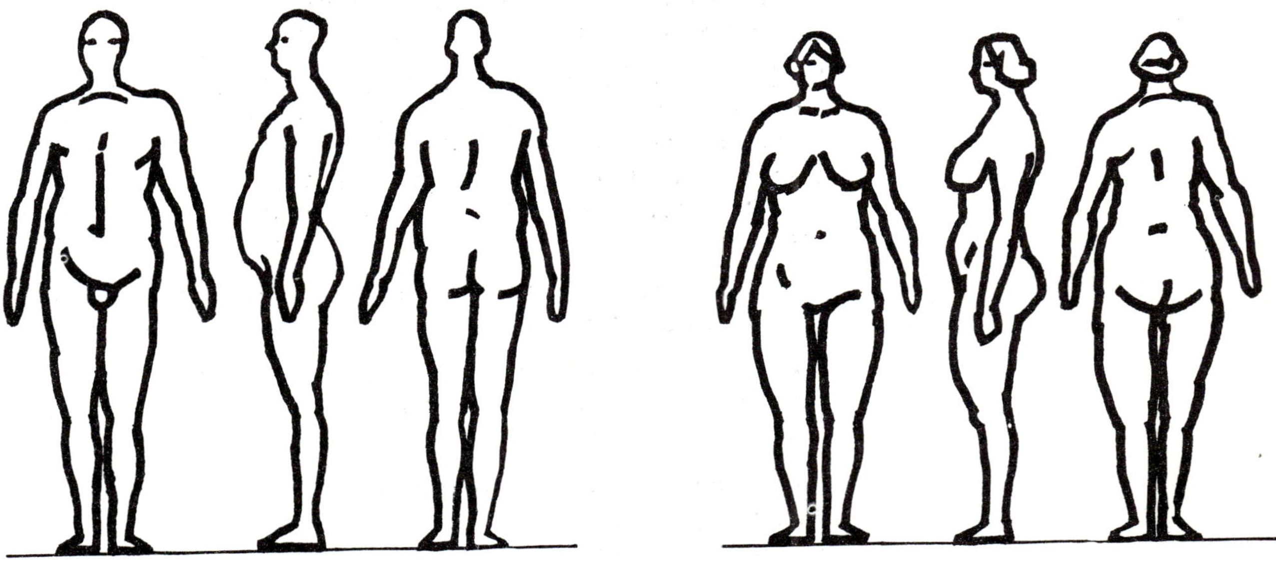

rating 6·3·2

likely to be found between particular physiques, than those between the average physique of two regions. It is difficult to define race if one wanted to, so much mixing of peoples has taken place and is being intensified. However, classification of race has no purpose in this book; in fact rather the opposite. Anthropometry and applied physical anthropology is concerned with occupational groups rather than any other groupings. Variation of trunk height to leg height, or the variation of body-builds, wherever they are found are measurable differences that we use on behalf of a population to design products for its use. Static and dynamic anthropometric data will provide the designer with an armature of dimensions around which his ideas can grow. Static anthropometric measures are made on the figure in standardized positions between fixed points. Unfortunately there is still no unanimity of opinion on the number or position of the fixed points. Care should therefore be taken to see that the anthropometric sequence chosen is suitable for the scheme in mind.

Dynamic anthropometric measures are taken on or around the figure when it is in any position other than the fixed ones. This covers the whole continued range of movement available to us at leisure or at work. They involve complicated measurement and in nearly every case must be determined within a framework of specific circumstance. Both types of measurement require anatomical knowledge so that the fixed points can be identified and communicated, and the range of body parts understood.

28

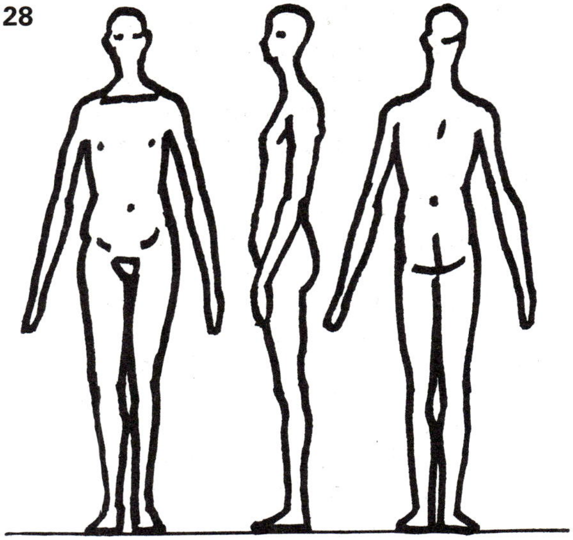

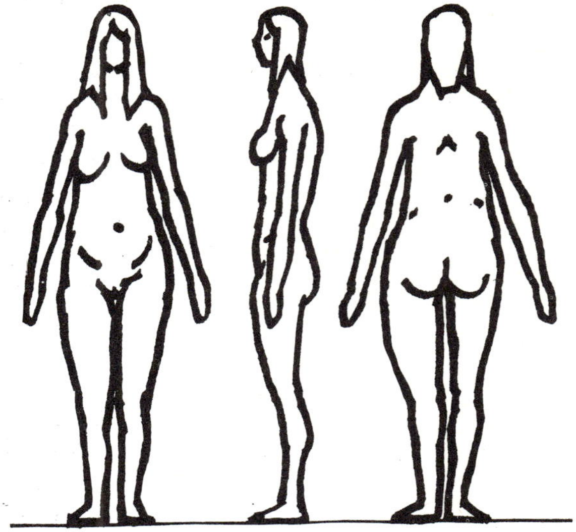

rating 4·1·5

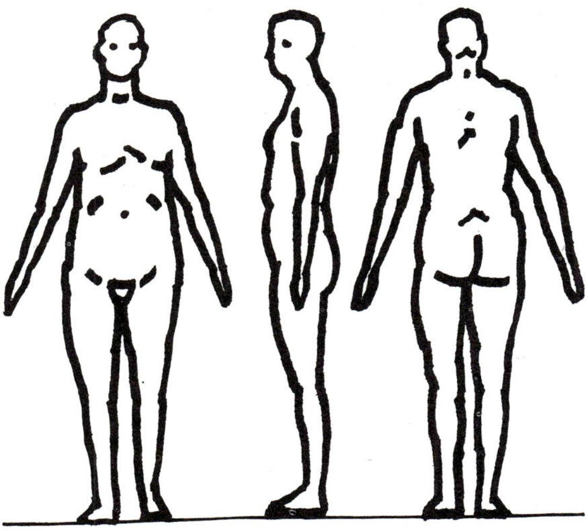

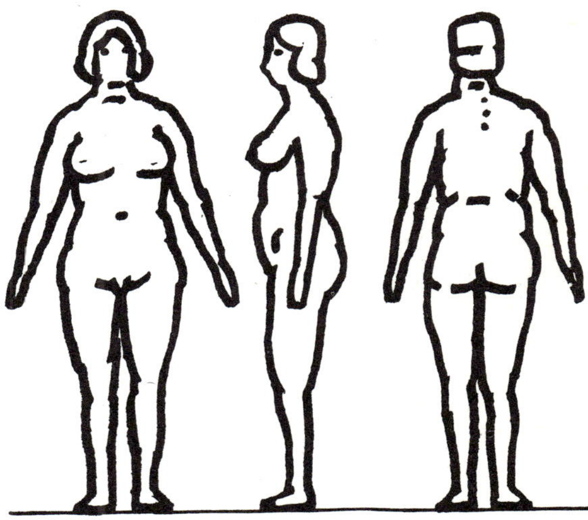

rating 5·4·2

29
Standing Posture for standing height. "A" represents an adjustable cross bar.

Practical Anthropometry

The human body is a very complicated form built up in a variety of solids, none of which has a very close approximation to the structures found in an introduction to solid geometry. It is on the other hand extremely simple to indulge in measuring the human body from a linear point of view and to go on increasing one's score in this field with very little profit. Therefore, if the figure is to be measured in such a way that it will yield useful results the measuring technique must be considered and a plan formulated to work by.

The figure must be unclothed. Measuring from 'fixed points' on the unclothed figure forms the bed-rock of all anthropometric work. From the unclothed figure we have derived a great deal of statistical information on the peculiarities of human physique and their distribution. For this part of our study we will assume the figure static in the sense that the subjects being studied are placed in two positions or postures. These two postures are simply those of standing and sitting. The standing posture is the normal position which is illustrated in the beginning of Section 1. (Illustration **1b**.) But the subject's back should be as upright as possible and against a flat vertical surface when statue or standing height is being measured. (Illustration **29**.) The sitting position of the subjects would be such that their back was also upright against a flat vertical surface, the seat height so arranged that the legs would hang unsupported (Illustration **30**). This would be the position for ascertaining trunk height or sitting height from the seat surface. With the feet supported will be some use of the muscles in the pelvic region to stabilise the trunk position. This although natural enough does not allow accurate measurement and so the lower legs are allowed to fall freely when trunk length is taken. For sitting height for industrial and seating purposes the feet must be supported at floor level, the weight of the upper part of the body being borne by the ischial tuberosities. (Illustration **30** and discussion on seating on pages 143 and 145.) Because the important relationship in this case is lower leg length, or seat height above ground or foot rest level. In both postures the eye level should be horizontal, so that the apex of the skull is always in a uniform position from subject to subject. These two postures are good for a great number of measurements required, although for anatomical details of the head, hands and feet or width of limbs, the subject may be arranged to the measurer's convenience.

For positions outside the normal, or standing or sitting ones which may be required for industrial reasons, the subjects must be arranged in specially constructed 'mock-ups' of the situations envisaged. This covers all the situations where the back of the subject or operator to be is not in an upright position and where the legs are not in the positions previously described. Any posture arranged for measuring must be clearly described by the surveyor, and all 'fixed points' or measuring points given a simple and correct description.

Subjects should be measured in the morning rather than the evening, because people are taller to the extent of a half inch or slightly less in the morning. By anthropometric tradition measurements for limbs are taken from the left side of the body. Weight should be taken on a well-serviced beam balance scale.

Standing or Standing Height

This measurement must be taken with the subject standing against a vertical background surface in a normal erect position, the shoulders, buttocks and heels lightly touching the background. Account must be taken of the figure type when the body surface is arranged against the background. The arrangement of the head, in a natural level position, with the eye horizontal, will however take care of the spinal adjustment of the subject to the normal standing posture. The measurement is taken from the top of the head to the ground; an adjustable cross bar being brought down to the head and the measure taken from a vertical scale. (Illustration **29**.) For basic anthropometry the subjects should be without shoes, and holding themselves erect to the best of their ability.

30
"A" represents an adjustable cross bar.

"**a**" represents the basic seated position from which all sitting postures for commercial or industrial use can be derived and measured.

"**b**" represents the seated position from which anatomical trunk length can be measured.

The Body: Trunk Length

This measurement is again taken against a vertical scale with an adjustable cross piece being brought down on to the top of the subject's head; the measurement being from the head to the horizontal sitting surface. As with the standing posture account must be taken of body physique and the posture controlled by some adjustment of the subject's head so that the gaze is horizontal and the trunk is as erect as possible. (Illustration **30**.)

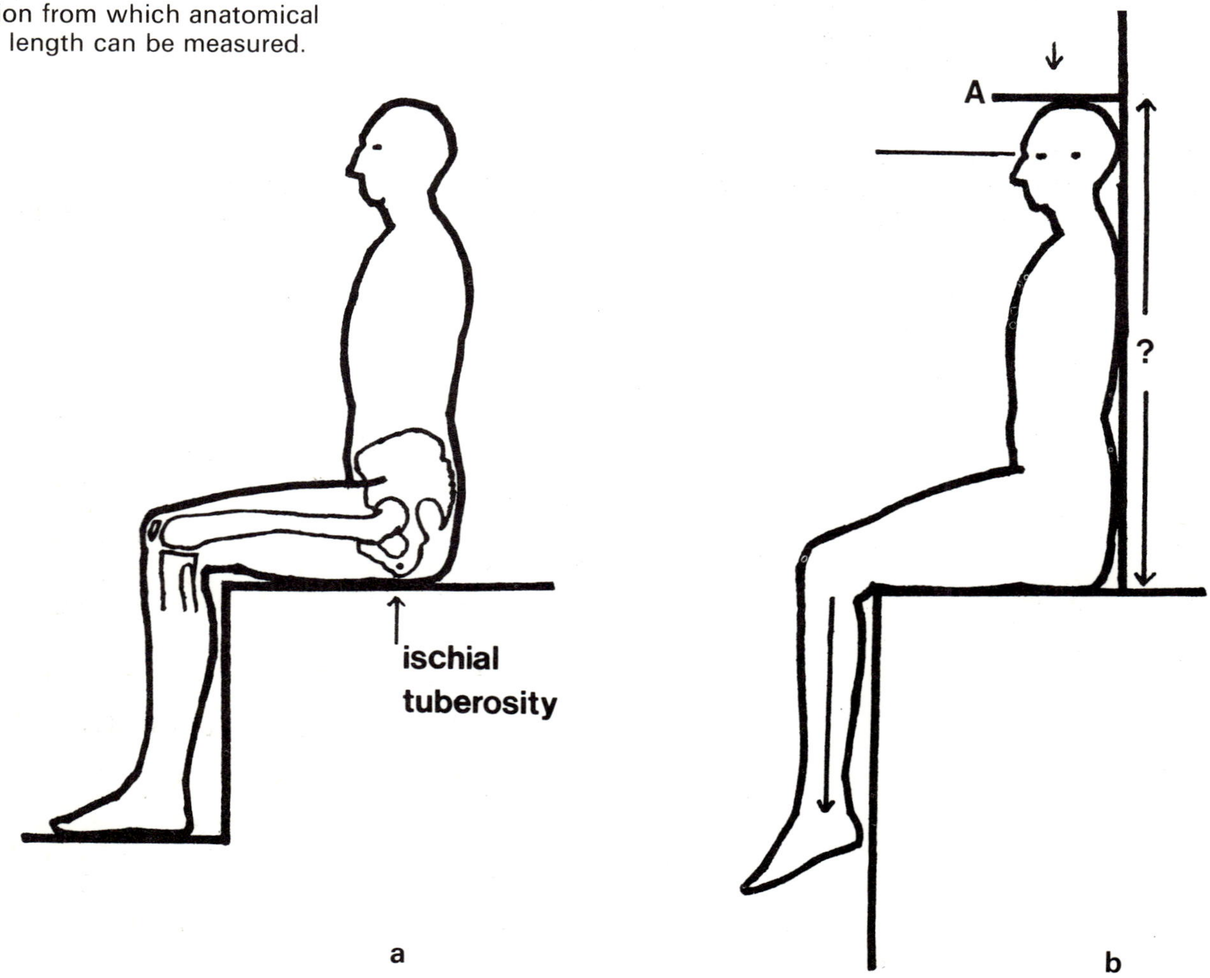

Trunk Girths or Circumferences

All measurements of the upper part of the trunk should take into account the changes of measurement caused by respiration and this includes measurement of the waist. Three movements in time can be considered; when the chest is in complete expiration or full inspiration, or in a judged half-way stage between the two preceding ones.

Measurements of girths or circumferences can be taken at various levels on the trunk. Some 'fixed points' for levels are armpit height or axillary chest circumference; nipple height or xiphoid cartilage height. (Illustration **31a**.) These measurements can be taken with a linear metal tape measure.

Waist Girth or Circumference

The subject should be posed in the normal position. The waist measurement will be the least girth measurement

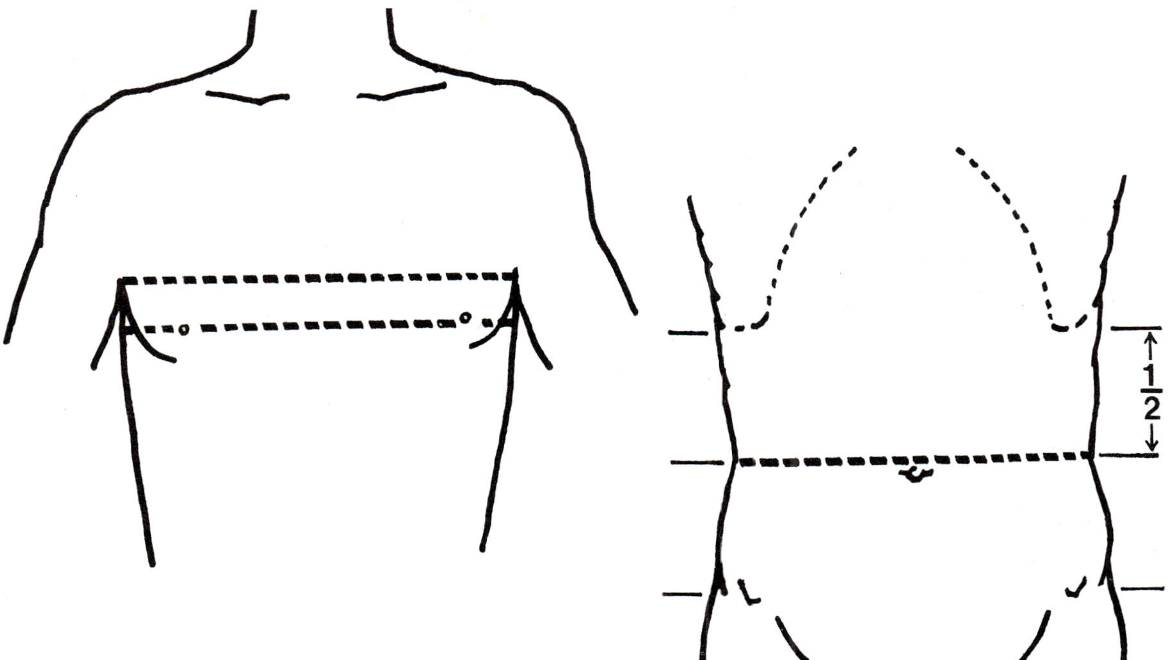

a

b

31a and b
Illustrations for chest and waist girth measurement.

recorded between the lower points of the sub-costal angle and the anterior-superior points of the iliac crest. (Illustration **31b**.) This measurement, as with all other girth measurements, is taken with a linear tape measure.

Hip Girth Circumference

This is taken with a tape at the great trochanters of the femur bones. (Illustration **32c**.)

Measurements of breadths can be taken on the trunk but they necessitate the use of an anthropometer. This measuring instrument, which is essentially a rod with two arms, one arm of which has a recording counter, requires practice in its use. The practice required is centred largely about the consideration of pressure of the anthropometric arms on the subject.

Trunk measurements include; chest depth; chest breadth; hip width and depth and shoulder widths.

The Thighs and Legs

Thigh Circumference

This measurement can be taken, but is extremely difficult to define in a uniform manner. The subject must stand in a fairly relaxed attitude, although not slumped, with the feet placed evenly on the ground. The tape must be arranged parallel to the ground just touching the lower edge of the buttocks.

Leg Length

Leg length is commonly taken as standing height less sitting height. This is easy but only a rough measure. The attempts at leg length measure must depend on the use to which the measure is to be put, for the top of the leg is difficult to determine. Two fixed points for the top of the leg could be the great trochanter, although this is a vague area rather than a point for a linear height measure, or the inferior most part of the ischial tuberosity. The crotch offers a third position used in clothing measurement.

Leg region measurements include: knee width; calf girth; foot length and breadth; ankle thickness and breadth.

The calf girth is taken as the maximum measure found when the measuring tape is at right angles to the shaft of the leg. The lower leg must hang free for this measurement.

The Shoulders

The width of the shoulders is taken from the most lateral point of one acromion process to the other. (Illustration **32a**.)

32a, b and c

a Shows lateral points of acromion processes in clear view enclosed by anthropometric rod.

b Shows by means of a plan view how most lateral points of acromion process are enclosed by tissue of the deltoid region.

c Shows position of great trochanters of femur bones and hip girth circumference measure.

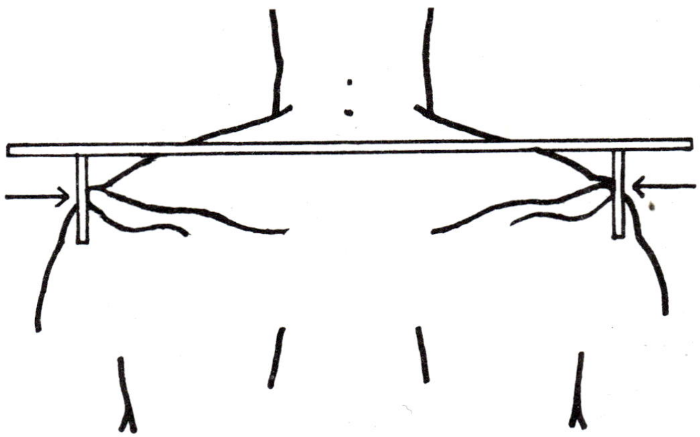

a

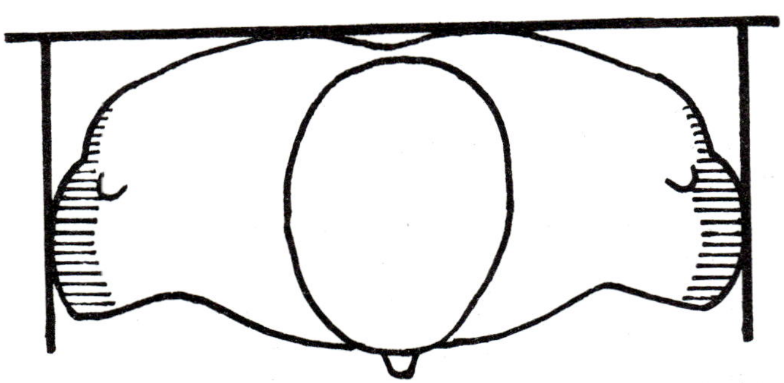

b

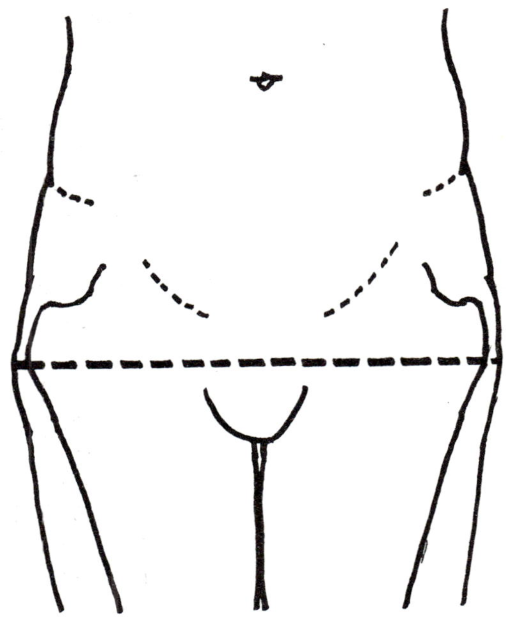

c

For industrial purposes this however leaves out of account post-deltoid development with or without subcutaneous fat. The bony width of the shoulders is taken with the adjustable anthropometer. Post-deltoid measurements might have to be taken with the subject seated upright and lightly against an horizontal scale with adjustable arms. (Illustration **32b**.)

The Arms

Arm length can be taken with the arms in the normal position. The length then being from the most lateral point of the acromion process to the tip of the third finger.

Arm span can be measured from third finger to third finger when the arms are fully extended.

Arm girth can be taken halfway between the elbow and the tip of the acromion process. Forearm girth can also be taken.

Arm region measurements include; elbow width; lower arm length; hand length and breadth; and wrist thickness and breadth.

The Head and Neck

Many forms of measurement, devised for the head, would be out of place here. From the designer's point of view their usefulness declines as their number increases. Head circumference can be taken with a metal tape.

This can be taken in a line connecting the two frontal eminences and following round the head from these. Other surface measures of the cranium can be devised to solve particular problems of millinery or uniform. (Illustration **33**.)

Head length and breadth can be taken with spreading calipers.

Neck Circumference

This can be taken at the level of the vertebrae prominens when the tape is resting above the head of the sternum at the front of the subject.

A large number of other measurements may be taken. A high proportion concern height from the ground and a number of others include measurements of length or width. Many of them will be found in terms of measurement dimensions in work spaces and living environments in Section IV.

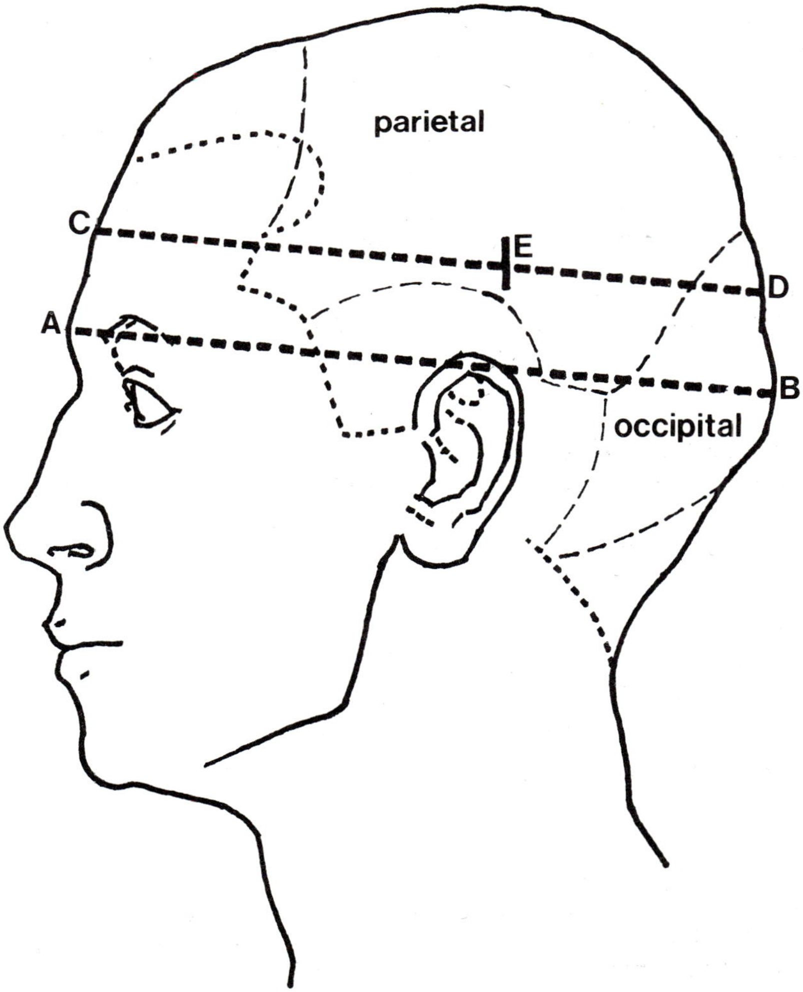

33
A.B. These points mark the setting for the maximum head circumference measure. It is taken from the glabella "A" between the eyebrows round to the furthest projection of the occipital bone "B" the opisthocranion.

C.D. Is another circumference measure taken from the frontal eminences "C" and passing round the parietal bones at their widest point "E" and round the back of the head from these "D". Points "A"."B". would give maximum head length with spreading calipers.

Points "E", one each side of the head, the euryon would give maximum head breadth with spreading calipers.

Glabella, opisthocranion and euryon are three of a number of landmark points on the head used in physical anthropolgy.

Section 3 **The Treatment of Measurement**

When man and his material needs are the objects of study it is more convenient to consider him *en masse* in terms of populations.

Populations may be surveyed and sampled as a whole or in specific sections of users. Populations can be defined fairly readily in relation to a design project under consideration. In fact the design project itself more often than not chooses the population. Dentists' chairs, school desks and cooking ovens pick their own victims.

Using simple standard statistical methods it is possible to establish useful facts about a chosen section of a population. Calculation of an average weight; or an indication of the range of weight by giving the lightest person and the heaviest person are two answers easily discovered.

The examination of a series of weights to divide it into classes of lightweights, medium weights or heavy weights is a third obtainable answer. Each answer in statistics depends on the manner in which the question is framed. In applied anthropometry questions about human measurements are prompted by design projects. The fact that the questions must be asked in terms of quantitative measurements is an initial assurance of success. Applied anthropometry excludes qualitative biological differences, e.g. different blood groupings or physiological changes with age. It also excludes simple numerical differences in anatomy, e.g. numbers of muscles or numbers of digits. Quantitative measurement does include alterations of statue, body structure and muscle power during a life-time. All measurable human quantities can be graded between two extremes, although the majority of measurements cluster somewhere in between the two extreme values, and with enough statistical evidence total range ratios can be calculated.

CHART 11 *Some Examples of Total Range Ratios*

Statue (adult)	1·25 : 1
Total arm length (adult)	1·5 : 1
Total body weight (adult)	2·5 : 1

It can be seen that the total range ratio of human measurements is comparatively narrow against the possible wide range of dimensions the designer can achieve when he creates a new design. (Illustration **34a**.) The lesson for the designer here is that, detailing, and the careful consideration of slight adjustments to measurements, when designing a product for human use, is all important.

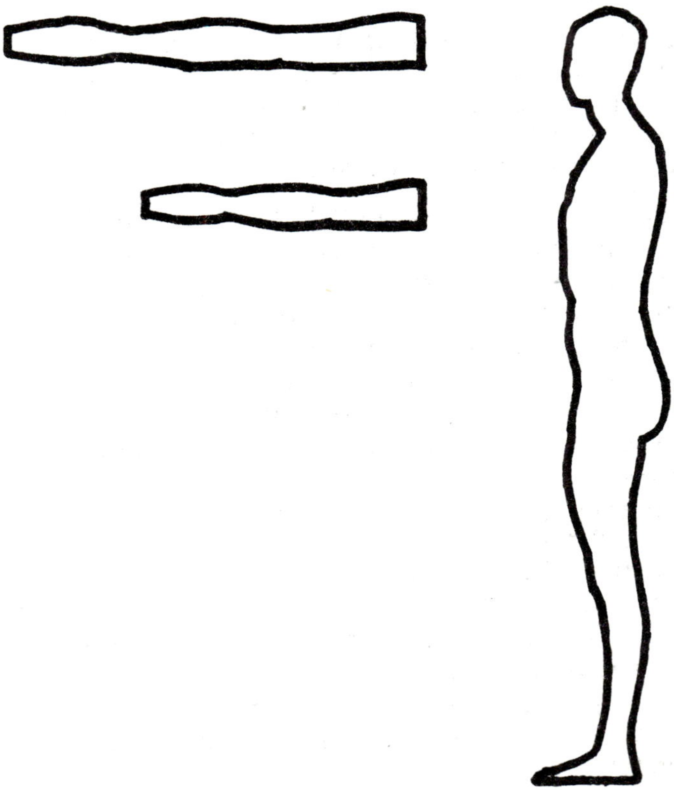

34a
Two examples of human range ratios in diagrammatic form.

Standing height 1·25 : 1

Total arm length 1·5 : 1

Careful thought must be given to the type of measurements taken from a population. Measurements must be relevant to the problem to be solved, e.g. in the case of a seated operator: is it lower arm- and hand-length or depth of trunk and thigh length that is wanted or both? (Illustration **34b**.)

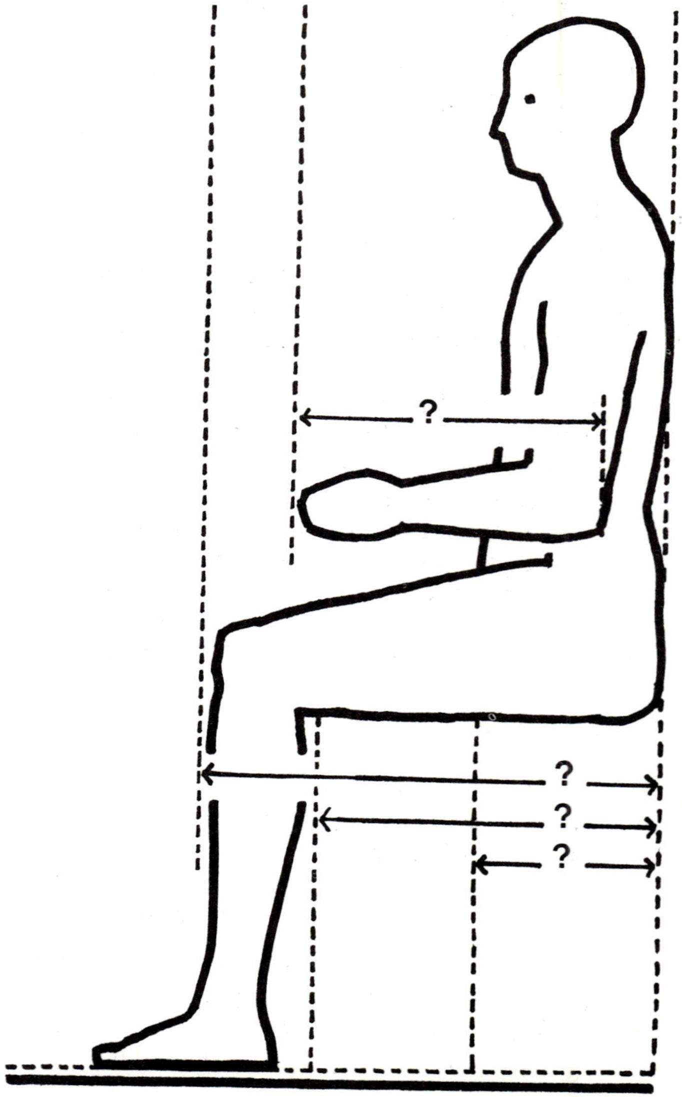

34b
Relevant measuring is important in two ways. It necessitates the formulation of a planned approach and the careful consideration of detailing. It also eliminates unnecessary measuring which wastes time and money.

It is therefore necessary to know the measurements or measurement differences required for the use of a particular pieces of equipment or garment. If it is found that a measurement can be invalidated from any point of view it is better to exclude it. When the type of measurement required has been decided it should be taken several times on each chosen subject. It is preferable for more than one person to take all the measurements. Although such comparisons are time consuming, they are well worth while. Errors must be reduced to a minimum so that each measurement is totally reliable.

When we define the scope and limits of a population for a particular purpose, sooner or later we find ourselves confronted with the average person. The average person is a concept of a typical person that emerges from the consideration of a number of persons, but it is nevertheless no one in particular. The average person is the general representative of a quantity of persons. It may be said that most men are taller than women, but of course some women are taller than some men. However, we can say the average man is taller than the average woman. In some fields of work simple everyday facts about a population that can be represented by an average person may be enough. A man rather than a woman; a factory worker rather than an office worker. To design for the average person, however, can lead to dangerous errors. Very few persons in a population are average in a large number of definitive measurements of bodily dimensions or capacities that could be examined in an anthropometric study. Percentages of a population may be inadequate in some dimensions or capacities and not in others. If we pursue the average in terms of more and more definitive characteristics we find that as the total number of definitive characteristics increases so the percentage of the 'average' person who can represent them all decreases.

The average person nevertheless remains a useful concept in statistical application. Anthropometry is the empirical study of man by definitive measurement. Statistical methods enable us to examine ranges of definitive measurements and make observations about their frequency of occurrence and distribution: statistical procedures also reveal significant trends and relationships in collected data that have direct application to design problems.

All the definitive measurements we can take on a person vary from most other similar dimensions measured on a number of other people. In fact we find that the variety of one measurement increases as the number of samples increase. Eventually we discover duplicates and so on, for the increase is not indefinite.

Two facts emerge from this statement: the range of human measurement is surely a matter of chance; and quite a large number of measurements have to be taken before we can have an understanding of their variation.

When a large sample of one kind of measurement of body dimension, say height, has been selected from a population a pattern can be observed in their distribution if they are listed in order of scale and number. If a graph line X is then plotted of the scale of the measurements against their frequency of occurrence we have as a result a symmetrical or bell-shaped looking curved line.

35
A normal curve.

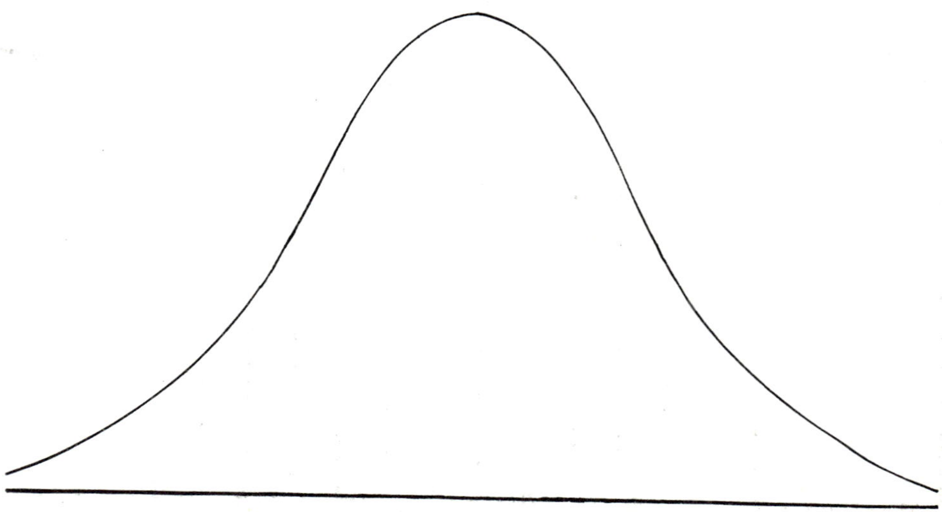

A curve produced in this manner from a quantity of varying measurements is called a Normal Curve or curve of Normal frequency distribution. (Illustration **35**.) This curve is one often met with in experimental biology where many continuously variable organic forms are being studied.

Because a distribution of this type lends itself so happily to mathematical treatment it forms the basis for much work in theoretical and applied statistics. It is so useful that if nature does produce a distribution which has some deviation from Normality, then appropriate transformations can be used to make it assume Normal form. Where factors are a matter of chance, as is the case with man's dimensional measurements, some measurements will appear more frequently than others. These measurements will obtain a higher score of repetition. The higher scores will tend to cluster round a centre or mean value. Other measurements of more extreme value and therefore rarer will be spread out on either side of the centre or mean value. (Illustration **36**.)

The centre or mean value, or the value which most frequently occurs, appears to be the average measurement, or the value of measurement which represents the average man. There are

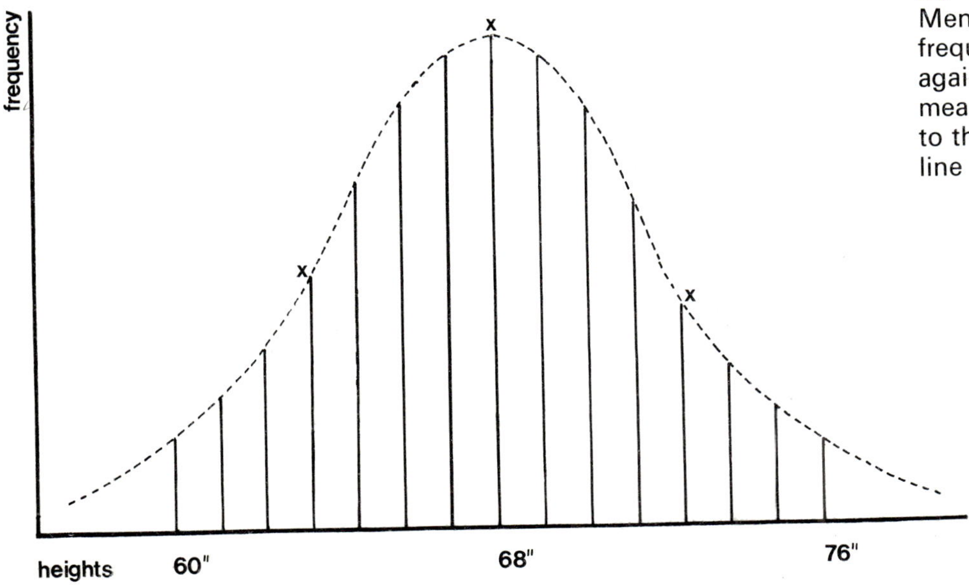

36
Men's height in inches; their frequency of number plotted against their range of measurement from the shortest to the tallest giving a graph line x.

however three different types of average which may or may not coincide; they are the mode, the median or the arithmetic mean. The mode is the value which most frequently occurs in a chance collection of measurements. So the mode is the representative of the commonly used average. To say that an average man is 5′ 8″ (1727 mm) in height is to say that more often than not a man is about that height. But it does not say that the mean of all men's height is 5′ 8″ (1727 mm). The median is the value that is the centre item in a chance collection of measurements. To say that the median of a range of men's heights is 5′ 8″ (1727 mm), is saying that 5′ 8″ is the value of the height half way through a number of values of men's heights arranged in order of scale.

The arithmetic mean is the measurement value found after all the values have been added together then divided by the total number of value occurrences. To say that the arithmetic mean of men's heights is 5′ 8″ is to quote the value given from a number of men's heights when added together and divided by the number of men measured.

In a perfect Normal curve distribution the three different types of the average coincide with one another. (Illustration **37**.)

37
Showing the mode, median and the arithmetic mean at the centre of a distribution. If a distribution is perfectly normal and produces a symmetrical curve all three forms of the average will coincide at the centre of the curve.

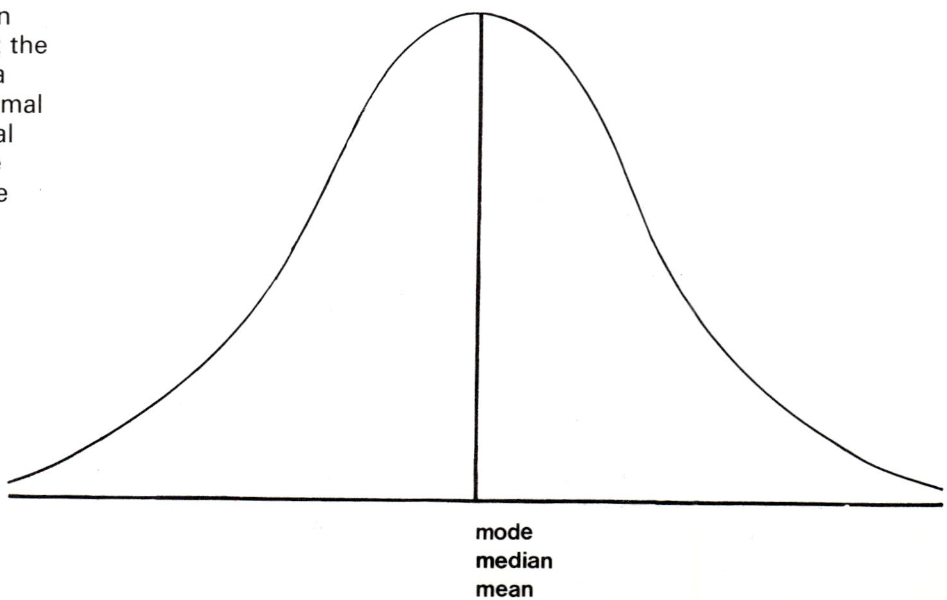

It is the interplay between genetic factors and environmental factors that determines the continuous variation of human measurements: determines, in fact, that their distributions are a matter of chance.

When we need to take a sample of measurements produced by chance, care and thought must be given to the manner in which we select the sample. To obtain a Normal curve that will yield useful results a sample of measurements must be both reasonably large and unselective. For our sample to be unselective it is necessary to devise procedures that are without bias, and that offer every subject of a population that is to be surveyed an equal chance of being chosen for measurement.

It is not within our scope to discuss advanced sampling procedures here, but they are recommended for further study. Thorough sampling necessitates collecting a sufficient quantity of data that will show the full ranges of particular measurements in a population. It is not possible or practical to use a whole population of users especially, if it is a large population. There is however a minimum number for a sample to be of any use. A hundred persons is the very least number

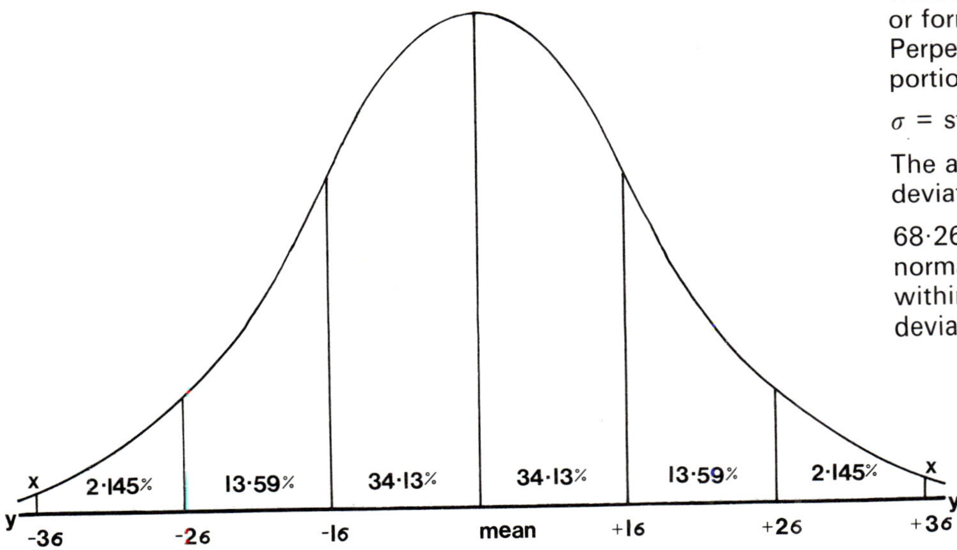

38
The area under a normal distribution divided by standard deviations.

Here x and y enclose an area or form a 100% unit. Perpendiculars enclose portions of the area.

σ = standard deviation

The areas enclosed by each deviation are shown.

68·26% of values in any normal distribution come within $\pm$ one standard deviation.

likely to produce a result that would be worthwhile. Three to five hundred is better, larger samples can be taken, but the collaboration of a team of investigators would then be required which requires organisation of a high order.

If very small samples only are obtainable assistance must be sought for statistical methods to establish their reliability.

In general the problems concerned with body extremities require smaller samples than do the larger body dimensions. The range of measurement of standing height or weight is much greater than the range of measurement for smaller anatomical details and consequently requires greater size samples.

The Normal curve is important: as we have already seen, because it is so convenient for the assembly and processing of new data, and therefore generally forms the basis for all statements of probability. This remains true despite a rather chequered history in mathematical history which dates back to 1733 when it first saw the light of day in print, as a result of the indefatigable study of Abraham de Moivre. If we take the area enclosed by the Normal curve as 1 or 100% we can see that smaller areas of it can predict probabilities in any binary event. (Illustration **38**.) De Moivre calculated a fraction of the total area under the Normal curve each side of the mean line to a distance of ± 1 which gave a value of 0·682688. ± 1 being one standard deviation each side of the centre line of the score.

If vertical lines are erected at ± 1 from the mean an area of 68·26% is enclosed or 34·13% for each deviation. Calculations for two standard deviations give values under x of ·2718 or 27·18% more area enclosed; 13·59% for each deviation.

Calculations for three standard deviations give values of ·04290 or 4·29% more are enclosed; 2·145% for each deviation.

The device of the standard deviation is a most useful statistic when considering the distribution of chance occurrences. When there is a Normal distribution we can say there is about a 33·3% chance of having a measurement outside one standard deviation, and about a 5% chance of having a measurement outside two standard deviations.

Standard deviations allow statements to be made about samples of measurement taken at random from a normal population. For example, if 120 measurements are taken of men's heights we know that 80 of the men's heights lie within $\pm$ one standard deviation, and that 114 lie within $\pm$ two standard deviations. That is if heights are Normally distributed; which in fact they are. By far the greatest amount of biological quantities are Normally distributed or very nearly so.

Standard deviations then can be taken as measure of dispersion round the average figure or the arithmetic mean. Any measurement in a distribution is the arithmetic mean of the measurements plus or minus the value of the number of standard deviations.

Often designing a product for a consumer population use means selecting to work in relation to values of measurement that come within the areas of one or more standard deviations. Two standard deviations contain the spread of the central 95% of values from the mode. This would be excellent coverage and in practice something rather less than this is achieved.

It is often useful to divide a distribution of measurements into groups or classes. (Example **1**.)

The number of classes must finally be decided from the quantity and variation of the measurements in a distribution. A number below ten might be used; but about 10 to 20 show good results.

When choosing the number of classes the number of measurements that each class contains is also being decided. If a distribution contains 400 measurements and it is divided into 20 classes then each class will contain 20 measurements. The number of classes chosen must finally be a matter of personal experience and judgement.

EXAMPLE 1 *Class groupings and frequency distributions*
Supposing a survey of men's standing heights had produced a distribution of measurements ranging from 58″ (1473 mm) to 78·8″ (2001 mm). The first step towards determining the number of class groupings is to find out the maximum quantity of measurements that could occur between these minimum and maximum values. 78·8 indicates tenths of an inch as the common value between measurements. That is 58″ (1473 mm) to 78·8″ (2001 mm) in tenths of an inch or 209 values. Tenths of an inch or 2·5 mm is a reasonable grading between measurements in anthropometric work, two fifths of an inch or (5 mm) being considered coarse. Classes are now fixed with the knowledge that between 10 and 20 show good results, the range and nature of the measurements being considered at the same time. A range of 209 is not very large, so a division by 20 would be too much. A division by 11 produces an exact result, so that all the classes are balanced within themselves, and 11 classes also brings about a useful increase of significance. The 11 classes would appear as follows:

mm's	inch's	
1473 – 1519	58″ to 59·8″	
1521·5 – 1566·5	59·9″ to 61·7″	
1569 – 1615	61·8″ to 63·6″	
1617·5 – 1663·5	63·7″ to 65·5″	
1666 – 1712	65·6″ to 67·4″	
1714·5 – 1760·5	67·5″ to 69·3″	←This class contains the mean
1763 – 1808	69·4″ to 71·2″	
1810·5 – 1856·5	71·3″ to 73·1″	
1859 – 1905	73·2″ to 75″	
1907·5 – 1952·5	75·1″ to 76·9″	
1956 – 2001	77″ to 78·8″	

The division of the data into 11 classes would make each class contain 19 of the one tenth or 2·5 mm increases in height. The difference in standing height through each class would be 1·9 inches or (48 mm).

The final step is to tally the actual measurements with their correct class grouping. Each measurement must be strictly placed within its class limits. The 58″ (1473 mm) measurement is scored into the first class with a tick; the 78·8″ (2001 mm) measurement is scored into the last class. A 75·6″ (1920 mm) measurement would be placed in the 75·1″ (1907·5 mm) to 76·9″ (1952·5 mm) class, and so on until all the measurements are tallied in.

58″ to 59·8″	I
59·9″ to 61·7″	
61·8″ to 63·6″	
63·7″ to 65·5″	
65·6″ to 67·4″	
67·5″ to 69·3″	
69·4″ to 71·2″	
71·3″ to 73·1″	
73·2″ to 75″	
75·1″ to 76·9″	I
77″ to 78·8″	I

The final scoring might look like this:

		Fc.
58″ to 59·8″	𝍸	5
59·9″ to 61·7″	𝍸 𝍸	10
61·8″ to 63·6″	𝍸 𝍸 𝍸	15
63·7″ to 65·5″	𝍸 𝍸 𝍸 𝍸	20
65·6″ to 67·4″	𝍸 𝍸 𝍸 𝍸 𝍸	25
67·5″ to 69·3″	𝍸 𝍸 𝍸 𝍸 𝍸 𝍸	30
69·4″ to 71·2″	𝍸 𝍸 𝍸 𝍸 𝍸	25
71·3″ to 73·1″	𝍸 𝍸 𝍸 𝍸	20
73·2″ to 75″	𝍸 𝍸 𝍸	15
75·1″ to 76·9″	𝍸 𝍸	10
77″ to 78·8″	𝍸	5

See fifth tallies are marked diagonally across the preceeding four. The tallies are then placed as a figure in a column on the right which is the frequency column.

The value of grouping according to frequency in classes is that it makes it possible for the mind to have an overall understanding of a random collection of figures and to place interpretations upon them.

It will be seen from the accompanying example that grouping measurement data into classes establishes ideas about data that could be gained in no other way, but at the same time the individuality of the measurements has been lost. It is in the consideration of the loss and gain of information that experience is required when choosing the number of classes for a distribution.

The construction of histograms is helpful as a visual aid when choosing classes in a distribution. Histograms show a series of rectangles for each class in a frequency distribution. A histogram is a graph which uses rectangles as building bricks forming steps. (Illustration **39**.) The width of each rectangle

39
A histogram constructed from the invented data of men's standing heights in the text.

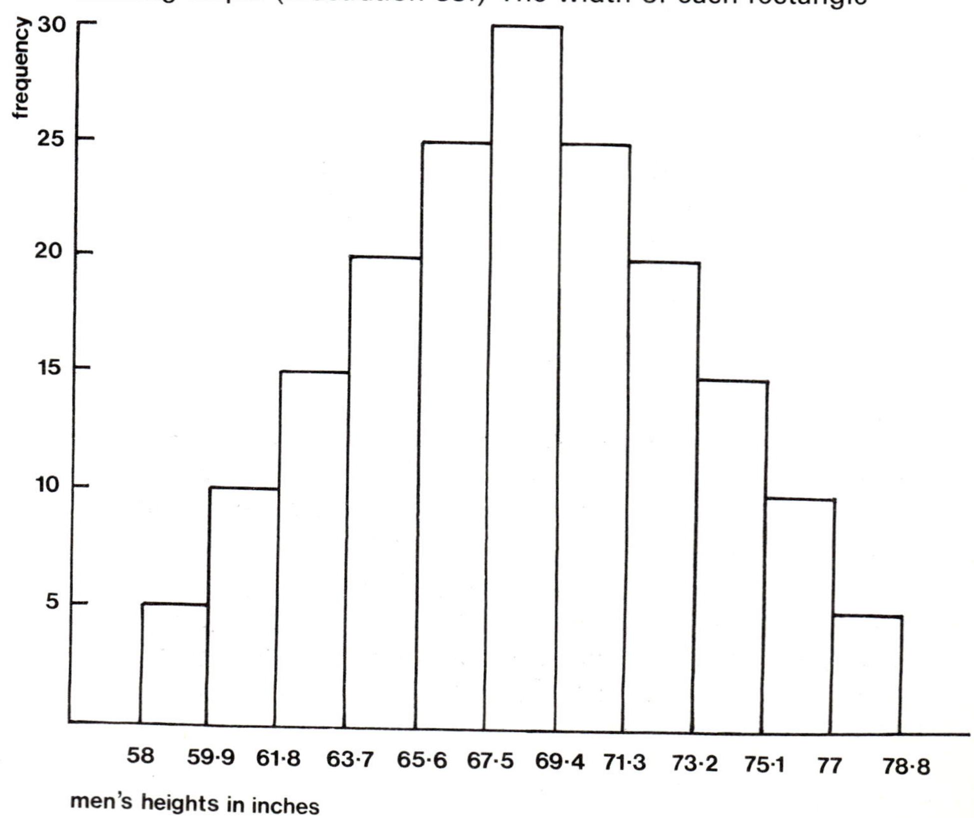

gives the limits of the measurements frequency of occurrence. The area of each rectangle is representative of the number of measurements in each class.

Histograms are therefore a useful means of determining the importance and significance of class groupings.

Another useful graph aid for studying the distribution of body measurements for design purposes is the plotting of a cumulative frequency distribution. (Example **2** Illustration **40**.)

EXAMPLE 2 *Construction of a cumulative frequency distribution*

To construct a cumulative frequency distribution add up the frequencies class by class from a frequency distribution. Using the material from the previous example we should have:

Frequency	*Cumulative Frequency*
5	5
10	15
15	30
20	50
25	75
30	105
25	130
20	150
15	165
10	175
5	180

Which would produce a total of 180 frequencies.

To construct a graph with the cumulative frequencies, the frequencies would be placed on a perpendicular axis from one to 180 and the classes of the distribution on the horizontal axis. The cumulative frequencies can then be plotted against the appropriate class limit. (Illustration **40**.)

Cumulative frequency distribution graphs will give cumulative percentages per population and offer percentile values as well. Percentile values are found by dividing any number of quantities into 100 equal groups in order of their size.

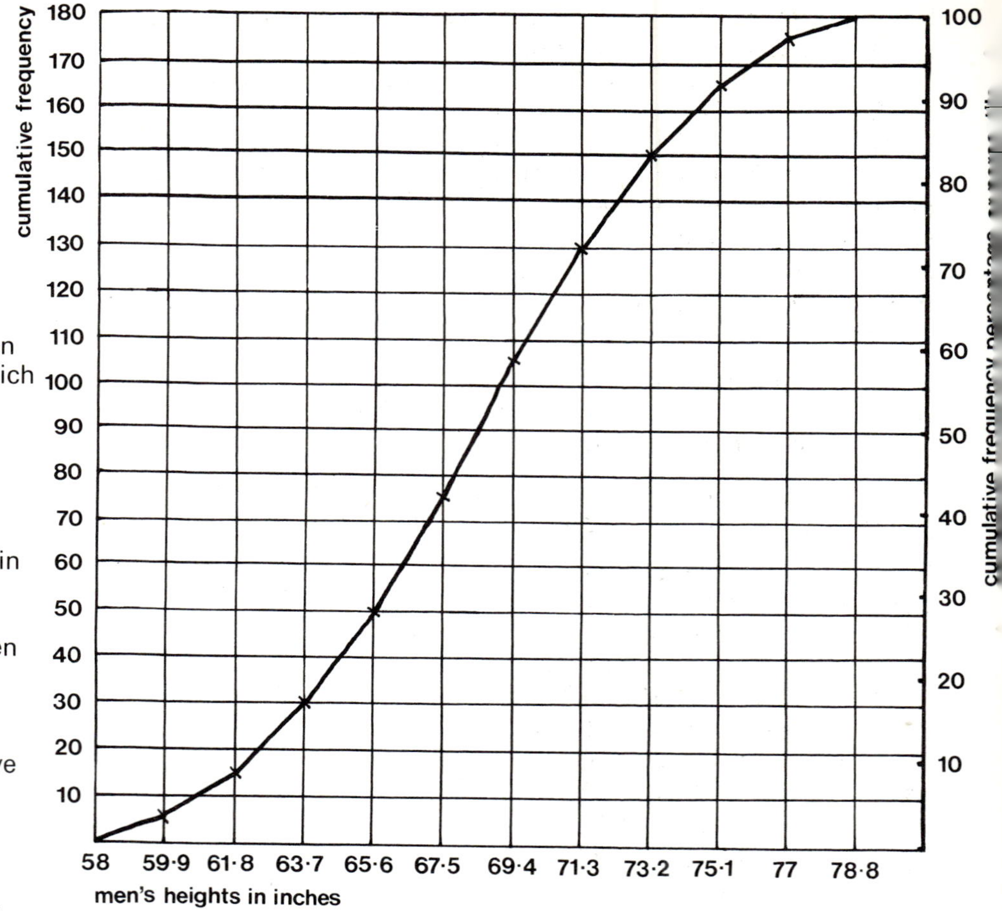

40
A cumulative frequency distribution. The right hand portion of the graph has been set off one extra division which is marked off in cumulative percentages or percentile values.

To establish the percentile value of a height: Decide where the height is situated in a particular class. Produce a perpendicular line from this point to the curve above, then read off from the right the percentile value which is horizontally opposite this curve intersection. Cumulative frequency curves are also called ogives. Cumulative distributions are useful and easy to consult for design purposes because they show percentages per population so clearly.

Percentiles

Using percentiles is another way of dividing a normal distribution from the least to the greatest in an orderly sequence. So extreme values of size; very small or very large, are represented by small or large percentile figures; the smaller or larger the percentile figure the slighter the chance of the occurrence of that value.

Percentile values are given in fractions of 100 in relation to an array of measurements. If the measurements in an array are ungrouped, finding the percentile merely necessitates counting up the number of measurements.

0 and 100th percentiles are the smallest and largest measurements in a distribution. If 58″ (1473 mm) is the smallest height and 78·8″ (2001 mm) is the largest height, the 0 percentile is 58″ (1473 mm) and the 100th at 78·8″ (2001 mm). The 10th percentile is therefore found by counting off from the lowest measurement one tenth of the measurements; the 20th percentile by counting off one twentieth of the measurements, and so on. If an array of measurements is grouped by frequency it will be necessary to use a process of interpolation. This method is described in Example 3.

Percentile values show the cumulative frequency of occurrence of measurements in order of their size as a percentage of the population being studied.

EXAMPLE 3 *To find a percentile value when data is grouped into classes in inches and millimetres.*

58″	1473	to 59·8″	1519	5	5
59·9″	1521·5	to 61·7″	1566·5	10	15
61·8″	1569	to 63·6″	1615	15	30
63·7″	1617·5	to 65·5″	1663·5	20	50
65·6″	1666	to 67·4″	1712	25	75
67·5″	1714·5	to 69·3″	1760·5	30	105
69·4″	1763	to 71·2″	1808	25	130
71·3″	1810·5	to 73·1″	1856·5	20	150
73·2″	1859	to 75″	1905	15	165
75·1″	1907·5	to 76·9″	1952·5	10	175
77″	1956	to 78·8″	2001	5	180

To find the 10th percentile

First find a tenth of 180 the total frequencies $\frac{1}{10} \times 180 = 18$

Count up the frequencies from the smallest end of the distribution until the 18th is reached.

The 18th comes within the third class with 15 frequencies.

We require three of the 15 frequencies to make 18.

To find the value of 1 frequency in this class divide 1·9″ (48 mm), the size of the class, by 15.

This equals ·126″ (3·2 mm).

So ·126″ × 3 added to 61·8″ (the smallest end of the third class) will give us the 10th percentile value.

·126″ × 3 = ·378″ (9·6 mm)

61·8″ (1569 mm) + ·378″ /9·6 mm) = 62·178″

Say 62·2″ (1578·6 mm) = 10th percentile.

From an array of height measurements we could find the value of height that would include 95% of the population. This information for example could be used to give the height for a door, that would only inconvenience the remaining 5%. A value for a 'reasonable fit' could be obtained for other products in a similar manner.

The dimensions for the design of products wherever possible should be derived initially from selected percentile values based on the distributions of users' measurements. In practice the attempt is always made to fit 90% of a population to a design, and try hard to incorporate 5% or 8% more.

The mean and standard deviations of a Normal distribution can be given in percentile values. (Illustrations **41a, b, c.**) Beyond one standard deviation percentile values can be seen to act as cut-off values to exclude the very smallest and largest size measurements which can rarely be fitted into a design scheme.

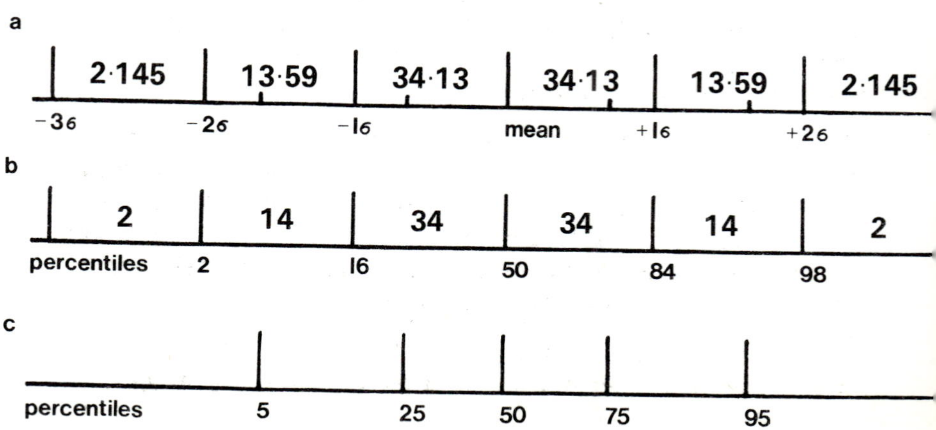

41abc
This diagram shows a comparison between a Normal distribution divided by standard deviations and a linear comparison of percentile values.

a represents the base of a Normal distritution and the intervals of standard deviations.

b represents a percentile scale marked with percentile values that relate very nearly to the ± deviations on **a**, the second percentile relating to -2σ, the sixteenth percentile relating to -1σ and so on. The percentage of values represented under the Normal curve between -1σ and -2σ 13·9% is very close to the fourteen percentile difference between the second and the sixteenth percentiles. Other percentile values can be compared in a similar manner. Comparisons between **a** and **b** is another way of demonstrating that percentile values give percentages of a distribution of a population. Percentile values can replace standard deviations, but are not so accurate.

c represents a percentile scale marked with well used percentile values and their relationship to ± 2 standard deviations is marked off on **a**.

One standard deviation each side of the mean value give percentile values of from 16 to 84. (Illustration **41b**.) So that the 5th and 95th percentiles become cut off values within ± 2 standard deviations.

Correlations

A correlated variation is the study of the relationship of two distributions when they have been combined together in the form of a table.

When two sets of variations are plotted together it can be seen whether they are independent of one another, or form a relationship by increasing or decreasing together or varying inversely with one another. By their use one can readily see the relationship of one variable with another. Correlation tables also give the intensity of the relationship of two measurements and are therefore most useful for anthropometric purposes in the clothing trade and in industry. Correlations are useful in the clothing trade because they can indicate common relationships of body measurements and the frequency of their occurrence.

Some body measurements are closely correlated, lengths of trunk and the arms and legs can be accurately predicted from height. Girths of the trunk and the arms and legs can be predicted from weight. There is also a high correlation between leg and arm lengths. As there are in existence a number of correlations of body measures it is worth considering before conducting a survey what is the minimum number of measurements to be taken.

However, not all correlations found in anthropometric tables may be safe to use for a new project. The body can be moved into a variety of positions to accomplish a task, and so it may be necessary to sample users in a 'mock-up' of a new design to establish what fresh measurements need to be taken; and which existing correlations can be utilized.

Correlations of new sets of measurements with already established anthropometric data may produce useful results.

EXAMPLE 4

By plotting men's crotch heights against their standing heights the intensity of their association could be studied. If both the distributions on the correlation chart were Normal the greatest intensity of correlation would be about the area where the more central values of the distributions met. The totals in the borders would give the frequency of distribution of the measurements. This correlation would have a tendency

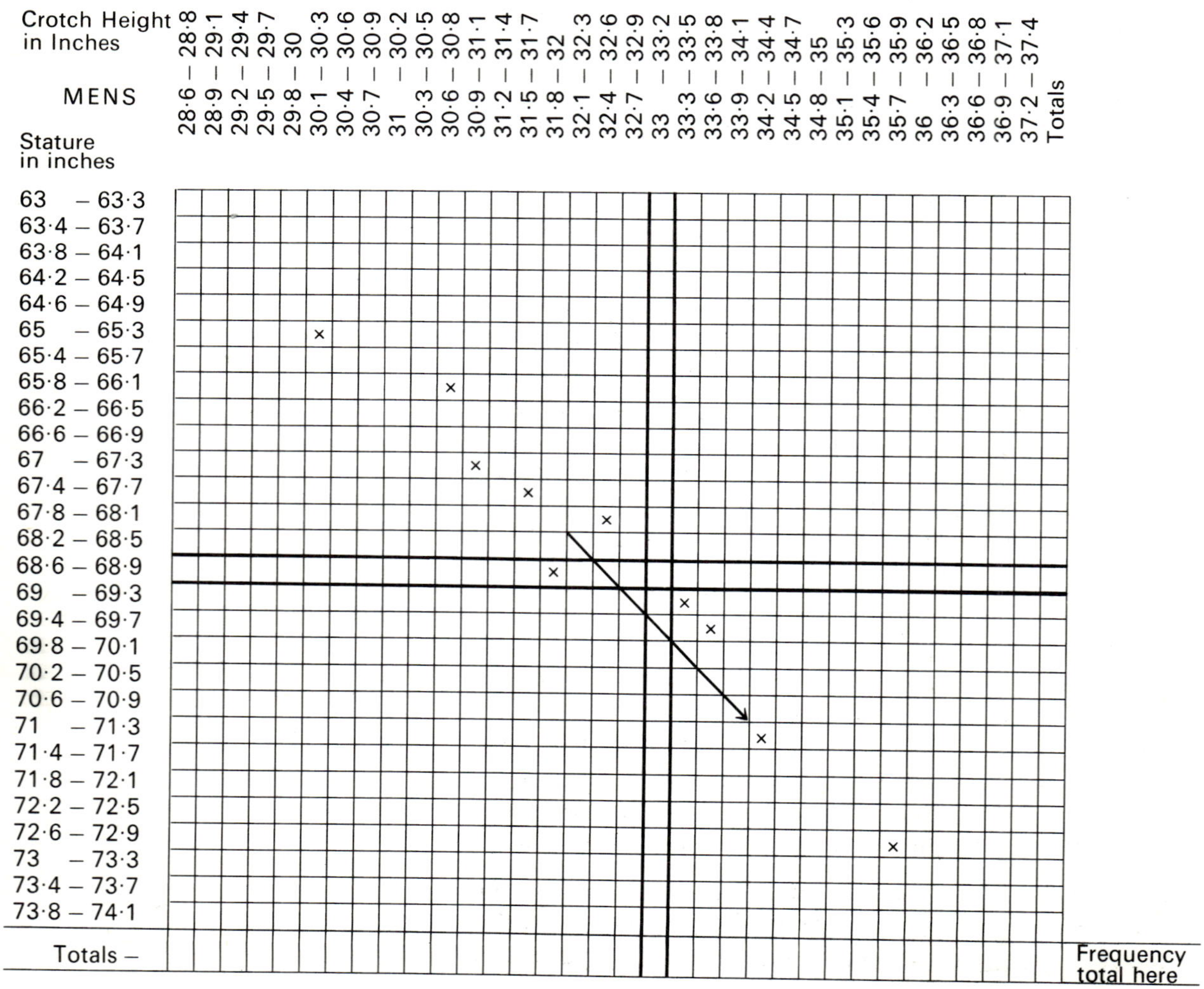

Frequency distribution. The darker lines give central values of the variables

to scatter about a diagonal direction from top left of the table to bottom right. This diagonal direction shows that the correlation is linear. Many relationships are of this kind. If the tendency of scatter is curved, mathematical transformations can be used to remove the curvature. Both linear and curvilinear relationships are produced by lines of 'best-fit': that is lines that pass through the points of greatest intensity of correlation. In statistical research it is important that the lines of 'best-fit' are very accurate and again mathematical treatment is called for. Such mathematical treatment can be studied in a manual of statistics.

The crosses on the accompanying diagram show how the intensity of correlation might build up. The arrow shows the expected diagonal direction they would make. The more distinct the diagonal pattern, the more close the relationships of the two variables.

Conclusion

When data is organised into tables or graphs they must not be too overcrowded. Other data from previous surveys for the same object should be organised in an equally clear and simple manner so that the whole can be compared. This editing of earlier work of a similar nature should be undertaken before one's survey is started. Measurements from a previous survey will assist in decisions about relevant measurements for a new survey.

Measurements taken for clothing purposes require much rechecking, for two people so often disagree about a single measure. As in all measuring, a group of the population being studied should be as widely representative as possible, the selection covering the full range of physiques. If clothing sizing is going to be successful measurement schemes must be related to the distribution of bodybuilds found by Sheldon.

All anthropometric sampling should be considered as research leading to the positioning or clothing of a person in one of two situations, because all conceivable environments, when considered in relation to the possible physical activity of its users, fit into one or the other of two main groupings.

One grouping will include all those objects or coverings of everyday use and wear. This grouping covers all access ways and spaces, furniture and architectural features, dimensions and clothing, etc.

The other grouping will include only equipment and clothing for highly specialised use and needed by a minority of a population. This grouping covers all special operating situations in industrial or commercial enterprises, in communications and endeavours in human aid.

It follows that for general use any design which imposes limitations by using unsuitable dimensions unnecessarily complicates its use.

It is for this reason that sampling for general use must seek for ranges of measurements wide enough to indicate values able to accommodate 90% of a population or more.

Indices

When we examine the shapes and proportions of the figures of a population we find that their characteristics can be more easily conveyed by combinations of measurements than by single measurements. These combinations we call indices. Indices state in the form of a percentage the relationship between two measurements. Indices are determined by multiplying one measurement of a bodily dimension by 100 and dividing the result by the measurement of another dimension. For example, an index which has been very widely used in physical anthropology is the cephalic index. It gives the relationship between head length and head breadth:

$$\text{Cephalic index} = \frac{\text{Max. breadth of skull} \times 100}{\text{Max. length of skull}}$$

If the result gives a figure below 75% the head is said to be dolichocephalic

a figure between 75–80% Mesocephalic
a figure above 80% Brachycephalic

Dolichocephalic is a type of head that is longer than it is broad. Mesocephalic a broader headed type. Brachycephalic a broadness of head which is verging on roundness.

Two other examples follow:

The cormic index gives the relationship between standing height and sitting height.

$$\text{The cormic index} = \frac{\text{seated height} \times 100}{\text{standing height}}$$

An answer of 45% would indicate a short trunk to leg length and of 55% a longer trunk to relatively shorter legs.

The thoracic index gives the relationship between two dimensions of the thorax. It varies naturally with body build, but more especially with sex and age.

$$\text{The thoracic index} = \frac{\text{lateral diameter of thorax} \times 100}{\text{anterior-posterior diameter of thorax}}$$

The adult thoracic index ranges between 130% and 135%

A large number of other indices exist but it is not necessary to list them all here because they have only the most marginal application to design. Once their function of imparting an idea of relative proportions is understood they can readily be invented when necessary.

Constitutional indices

Many attempts have been made to contrive indices that will indicate levels of bodily health and composition. They have not proved to be entirely adequate or reliable. But when one considers the fluctuating ranges of the many parts of man's physiological system this is not to be wondered at. To compile a constitutional index is like attempting to get a quart into a pint pot. The following is an example of a constitutional index to indicate underweight or overweight. The inventor was the German anthropologist F. H. Lorentz.

Age in years

2·5 to 6	Height-weight-(height-123) × ·7 = 100
6 to 14	Height-weight-(height-125) × ·5 = 100
14 to 18	Height-(weight-10) = 100
18 and above	Height-Weight-(height-150) × ·25 = 100

Where the height is in centimetres and the weight in kilograms

Over 100 indicates underweight

Under 100 indicates overweight

Section 4 **Dynamic Anthropometry**

Introduction
When undertaking the design of a piece of equipment or furniture or any environment for work or leisure we are always seeking an ideal or optimum situation for a person or persons. The ideal or optimum will be relative to a particular situation because, for example, what is 'comfortable' for leisure sitting, may not be seen in quite the same way when we are looking for a comfortable seated position which will at the same time allow maximum of effort to be exerted by a limb for a certain period of time, with the minimum of fatigue. So optimum for comfort might not always be able to be arranged to coincide with optimum for effort. What we can hope to do is so to design equipment, or space, that people using it are as well suited as the situation will permit and their tasks demand. This requires a great deal of imagination in the use of facts and figures about people. Man has limitations and many possibilities: not least in the last respect, an ability to stand up to stress. The designer must know that there is a price every user has to pay eventually for going beyond his capacity. The body is made of deformable materials, and man has very little instinctive ability left to tell him when his physiological processes have had enough. Undue stress on the body must therefore be curtailed. There is, as always, the other side of the coin. Too much rest and comfort lead to boredom. So that although undue physical trial should be eliminated, some evaluated system needing manipulation and selection should be left to enable the brain to be exercised. Man at his best can make new relationships visually or in the form of mental calculations, invent new combinations of control, initiate new actions, and react to slight changes in range sensory stimuli. He can also perform very delicate adjustments. These things may still be left to him, and other tasks may be passed onto machinery.

The construction of manikins can be recommended for the initial stages of a design. If they are accurately measured and constructed they give the facility of an efficiently presented set of inter-related measurements. They can be moved easily over the drawing board and can supply some speedy answers. However, they can only give information about a problem from a rigid, two-dimensional point of view. The designer must take his enquiries a great deal further if he is to satisfy the requirements of a live operator making dynamic movement in three dimensions.

During the initial design stage as much use as possible should be made of all relative and existing data. Engineers' and other designers' findings are relevant, as are comments from users and manufacturers. A continual return to the drawing board should be made to set each piece of new information into its place alongside the original draft. Other solutions of a similar design-problem should also be seen, and they should be probed for any weaknesses.

With a comprehensive set of data, that will accommodate a normal range of operators, set out on the designer's drawings, trials must then be carried out. This necessitates a 'mock-up' being constructed of the design. This must be considered a very vital part of a designer's job. If in the trials people have to be used to simulate operators or users great care should be taken that their physique is typical of the population of users of the situation one is constructing.

Trials bring to the forefront detailed consideration of allowances for ease of movement, vision or operation and they will always pinpoint the special conditions of particular cases. The extremes of the range of users should be carefully examined against specific operations.

Designing for human users is largely problem-solving in a physical environment. By trials we can adapt the environment and finally mould it successfully into an optimum envelope for use. The dynamic movements of a typical group of real users give final meaning or the lie to sets of anthropometric measurements. Imaginative and theoretically correct solutions can be vindicated by trials. The trial stage itself must be conducted rigorously by the designer to discover any weaknesses that are dormant. Improbable, but possible, combinations of movement and use must be examined. As improbable combinations of movements can often be the cause of accidents it is of great importance that they should be eliminated. Their probability and the extent to which they can weaken a system should be probed with practical tests.

If there are problems of use against time or force or of accuracy, recorded observations will suggest necessary alterations. If pre-packaged equipment must be introduced into a new situation the dimensions of movement and force required to operate it must be carefully examined so that it can be 'harmonised' with other equipment about it. It may have to be given pride of place in the user's cone of vision or easy range of reach.

The introduction of pre-packaged equipment brings to the forefront the problems of maintenance access. If maintenance is to be carried out by the user it needs special consideration in the initial design stage. Any form of maintenance of equipment, however, should be considered when a list of requirements for the new design are being undertaken.

General design considerations

Study the purpose and the operating performance details of the equipment, furniture, appliance, mechanical device or housing. Study the design solutions already in existence.

Dimensions and space

Allow for clothing, access, and movements in and about the work space. Attempt to accommodate between 5th and 95th percentile of users, or alternatively specific users. Consider any maintenance problems.

Operating conditions

Wherever possible use adjustable equipment. If users cannot be specified planning should consider women's percentile measurements rather than men's. If design is for non-specified users and there are any possible side effects from use, or difficulties or operation, they should be reduced to a level that could be tolerated by the older age range.

To facilitate speed
Consider the correct grouping and positioning of controls, and the grouping of indicators within the normal cone of vision. For average use limits on distance will be set by the lesser body measurements.

To facilitate precision
Consider support for body or limb segments and the operating position.

To facilitate pressure or force
Consider the amount of pressure or force and whether whole or part of a limb must be used. Consider body or limb supports and space necessary for operator to make adjustments of position.

To offset fatigue
Consider grouping of controls and the operating position in detail.

Trials
Make scale models and test. Use a full scale 'mock-up' of design. Check that the trial subjects are adequately representative of the future users; particularly that they represent the low and high percentile values of users. Check all detailing thoroughly. Determine all the limiting factors of the design, and ascertain by experiment whether they detract from optimum performance.

Adaptation

Man reacts to his surroundings. His surroundings can affect his whole performance.

A number of changes in his environment can be tolerated. It is possible to adapt to higher temperatures than one is normally accustomed to. It is also possible to continue working when the average light intensity is decreased. In more special fields of human endeavour it is possible to adapt to a lower air pressure and also to travel at faster speeds. That is not to say that if these changes are major that sudden adaptation can take place. Most new design situations demand some process of adaptation although very slight, and generally they are very readily tolerated. There are two possible changes in a person's surroundings that are harmful to performance and cannot be tolerated: they are changes towards brighter lighting and louder noise. Both these things are equally distressing and prove equally impossible to adapt to.

Anatomical movement

The human skeleton consists of rather more than 200 bones. (Illustration **42**.) Bone is an extremely tough and hard substance, but it does exhibit remarkable qualities of plasticity. It is being continually renewed in the process of growth, and makes a contribution of its own to the physiological processes of the body. However, it is the supporting framework for the muscles and their actions. This second duty demands that the bones should be strong but not brittle, and that they should be resilient enough to be able to withstand quite large forces of stress and strain from an almost limitless series of directions. The evolution of the body's framework of bones has given the possibility of many combinations of movement, and it is unique amongst animals in the types of joints it can offer. By the co-ordinated use of muscles and ligaments a large part of the framework can be held rigid, whilst a small portion is in action. To alter the position of the body as a whole or in part, the bones of the skeleton also offer a well-planned system of levers. The human body is a structurally adaptable instrument which, in normal health, can execute the dictates of man's will and conduct the most complex movements with the minimum of conscious control. We have only to consider the series of combined movements used in an everyday occurrence like walking, and simultaneously rearranging objects being carried, to understand this.

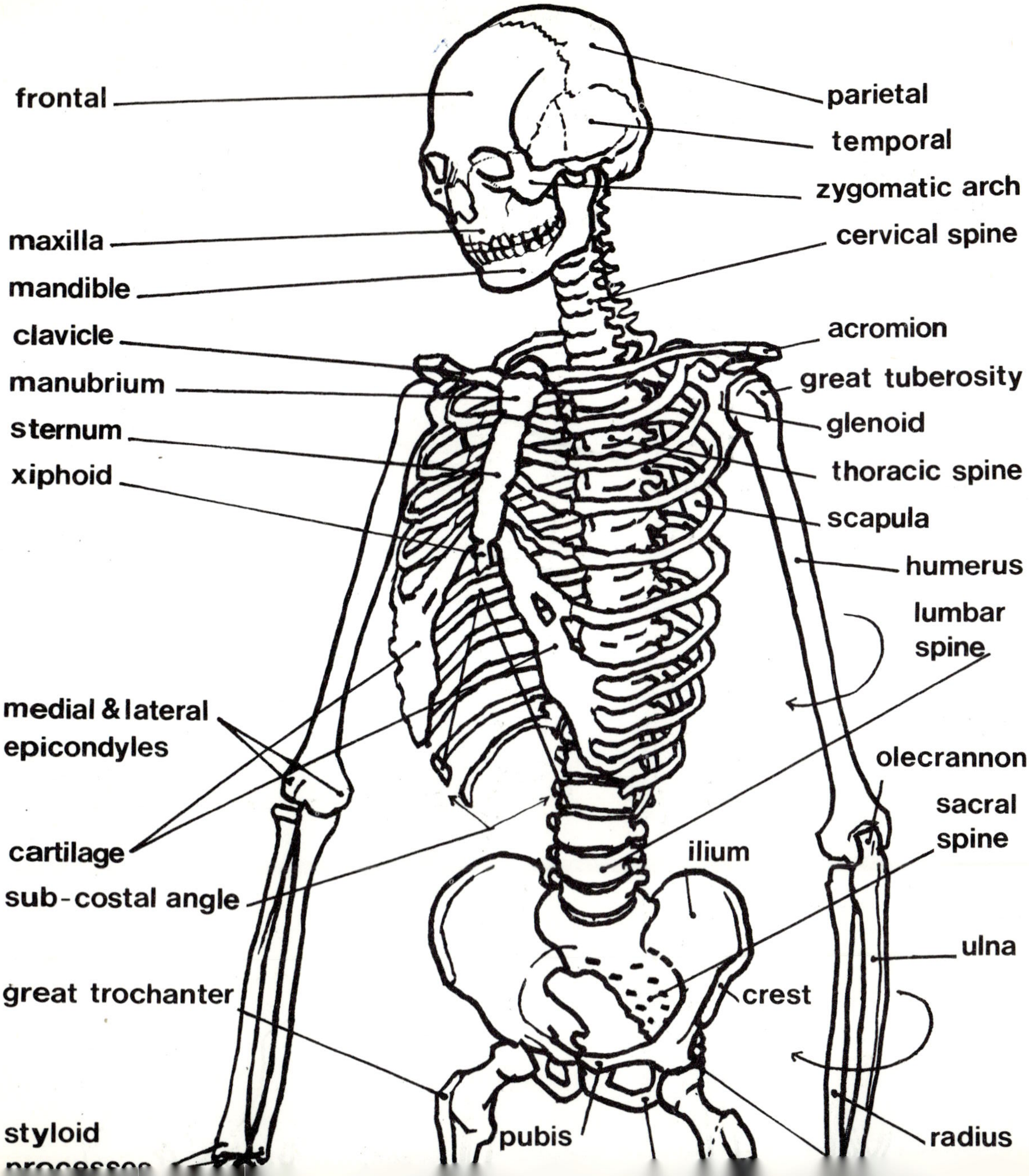
frontal
parietal
temporal
zygomatic arch
maxilla
cervical spine
mandible
clavicle
acromion
manubrium
great tuberosity
sternum
glenoid
thoracic spine
xiphoid
scapula
humerus
lumbar spine
medial & lateral epicondyles
olecrannon
sacral spine
cartilage
ilium
sub-costal angle
ulna
great trochanter
crest
styloid
pubis
radius

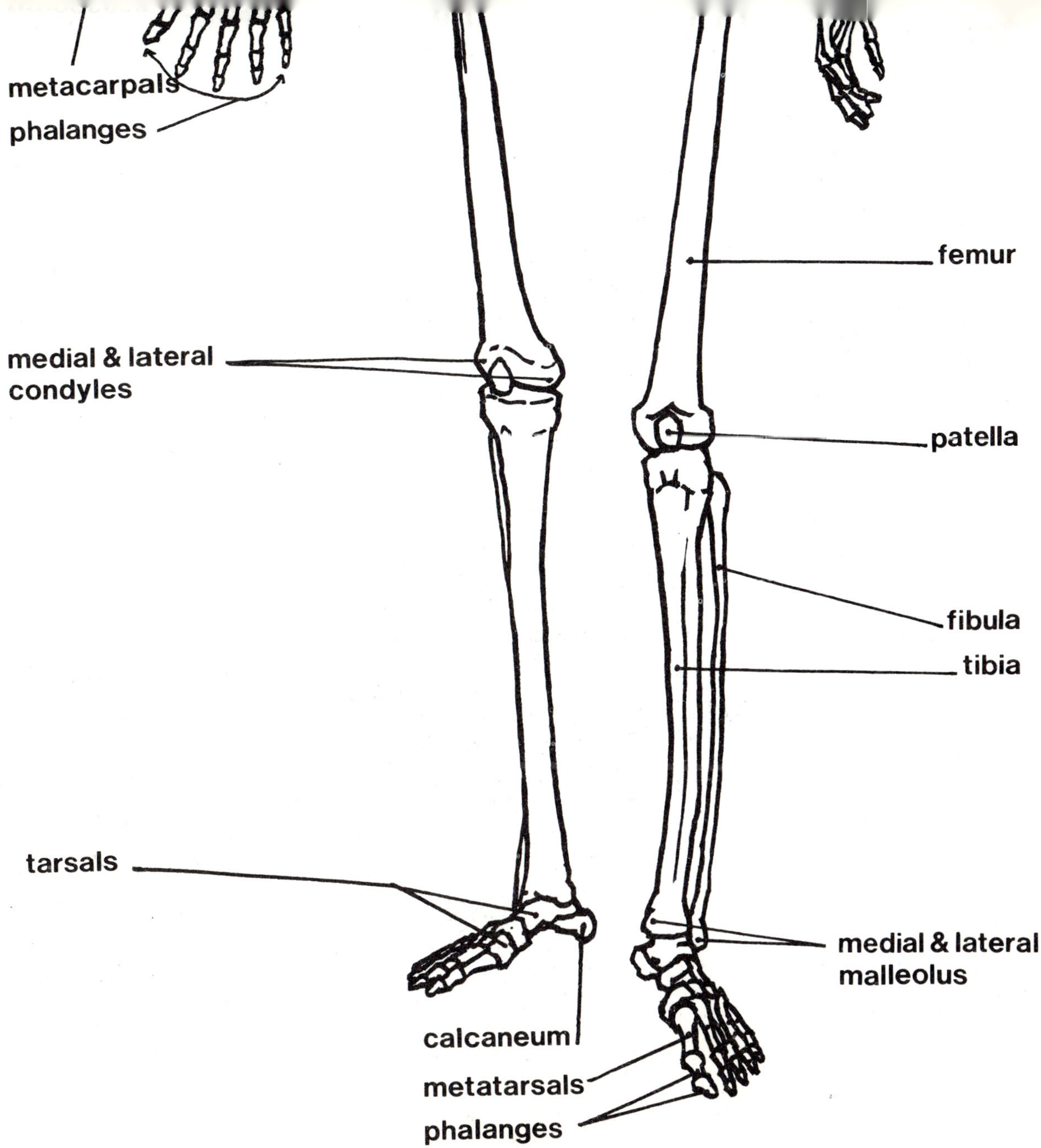
metacarpals
phalanges
femur
medial & lateral condyles
patella
fibula
tibia
tarsals
medial & lateral malleolus
calcaneum
metatarsals
phalanges

In the process of structural adaptation the architecture of bones has somehow been able to make use of civil engineering skills of a very high order. If we look at the femur bone of the thigh we can see that it has been tooled into a large number of subtly curved surfaces; and without any mathematical analysis we instinctively realise that the many facets of its surface reflect its mechanical purpose in locomotion and our upright stance. (Illustration **43**.) Like the femur, each bone of the skeleton is designed for a specific number of roles.

Man's skeleton shape is such that he can manage short periods of normal standing posture without much assistance from his muscles, and in this position he is relatively efficient. However, the state of equilibrium is an uneasy one. After standing for some time the weight of the upper body does threaten the knees with collapse, and when standing for lengthy periods we tense the muscles and ligaments of the leg joints to resist the pressure from above. These short bursts of muscle and ligament action means that standing for long periods necessitates the use of energy.

In assuming the upright stance man disregarded the advantages the quadruped has in the positioning of his thorax between four supports: the spinal column in this position acting as a true cantilever type bridge between the fore and aft of the animal. Man's thorax hangs from the spinal column. To solve the problem to some extent man has a shortened lumbar spine and an enlarged sacral spine. This allows a fair amount of turn at the waist without too much loss of stability. The skeleton is divided into separate bone formations. (Illustration **42**.) The whole bony formation is bound together by the more elastic connective tissues of cartilage and ligament. To establish a preliminary order for the skeleton we call the trunk and the skull the axial skeleton, and the limbs the appendicular skeleton. The trunk and skull are joined together by the vertebral column which by the nature of its construction of 29 separate bones, excluding the coccyx, provides a flexible link. The vertebral column as a whole consists of five distinct sections. The cervical spine or

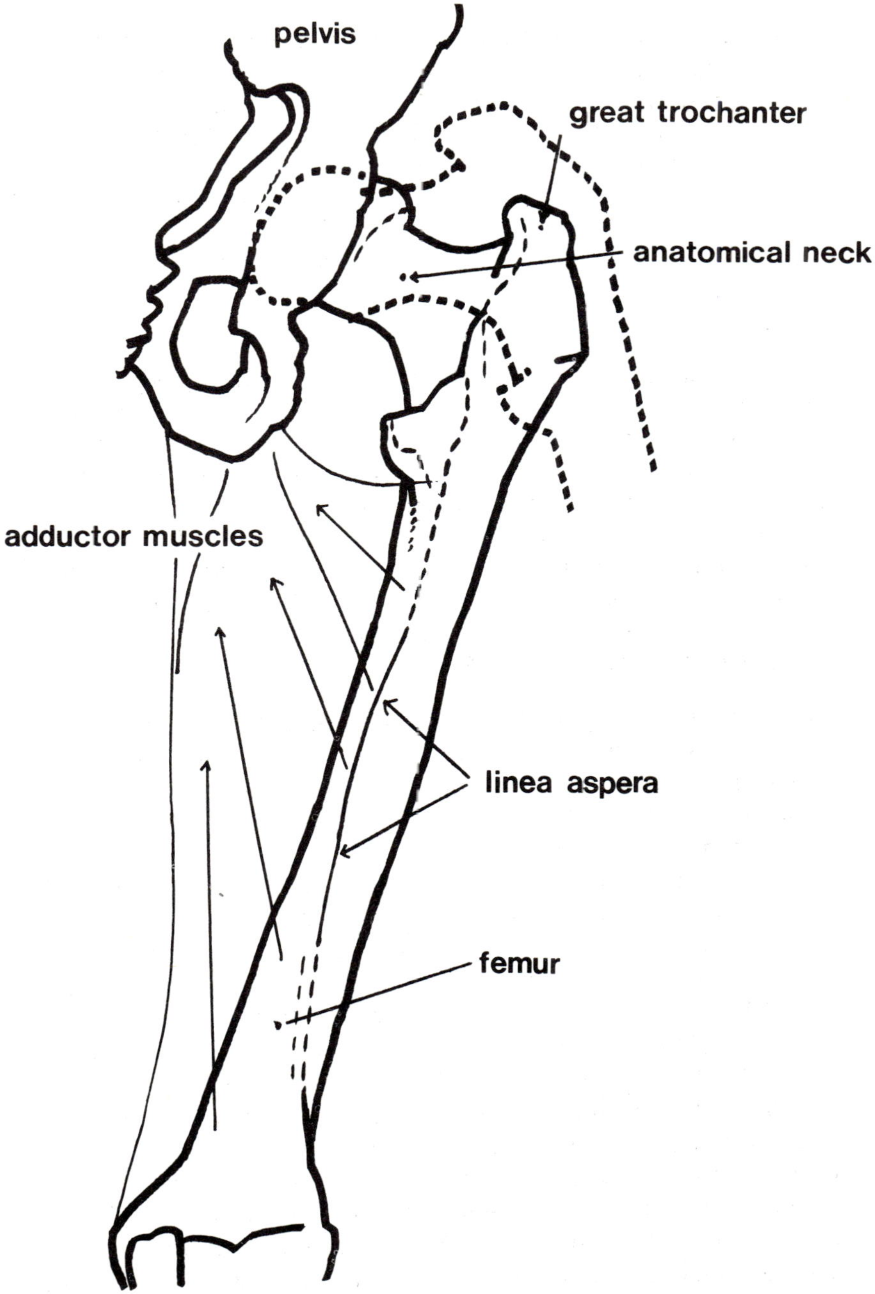

43
Diagram of the femur bone, posterior view. The linea aspera is a raised ridge on the bone. The dotted line shows the femur bone in abduction.

neck (seven pieces), the thoraccic spine (twelve pieces) and relating to the chest cavity, the lumbar spine (five pieces) relating to the waist, the saccral spine (five pieces) which is permanently joined and is the vertebral column's base in the pelvis, and the coccyx which variously contains three or four, or even more rarely five pieces and which is considered as the rudiments of the tail of our ancestors. (61% of people have four pieces.) The column allows us certain axis of movements. The cervical part allows up to 90° of movement in bending and the lumbar part up to 30° of movement in bending, the cervical spine allows rotation on its upper end of up to 55° rotation from the medial line, and the seven bones in combination allow 40° unilateral bending and between 45° and 50° in flexion and hyperextension in the sagittal plane. The thoracic and saccral spines are curved posteriorly and are present at birth. The cervical spine is convex-anteriorly and is fashioned when the six to eight months old human sits up; the lumbar spine is also convex anteriorly but is obtained slightly later at about 12 to 14 months when walking has been achieved.

The skull is considered anatomically in two parts; the bones of the cranium and the bones of the face. The cranium contains and protects the brain, the face contains the organs of recognition and communication and the palate.

The thorax contains and protects the lungs and the heart. Man's thorax is proportionally broader than that of other animals and it is also somewhat flattened in its anterior and posterior aspects. The greatest transverse section is in the lower half. The upper section is tapering in construction, which allows space for the movement for the four joints at the shoulder girdle from which the arm springs. (Illustration **44**.) Although rigid enough to support arm movements the bones of the thorax cage move in respiration.

The thoracic cage can be moved independently from the pelvis in the region of the lumbar spine and can in this way improve average reach distances by up to 16 inches (406 mm).

The pelvis or pelvic bones form together an open rimmed bowl-like structure. It supports and contains the abdominal

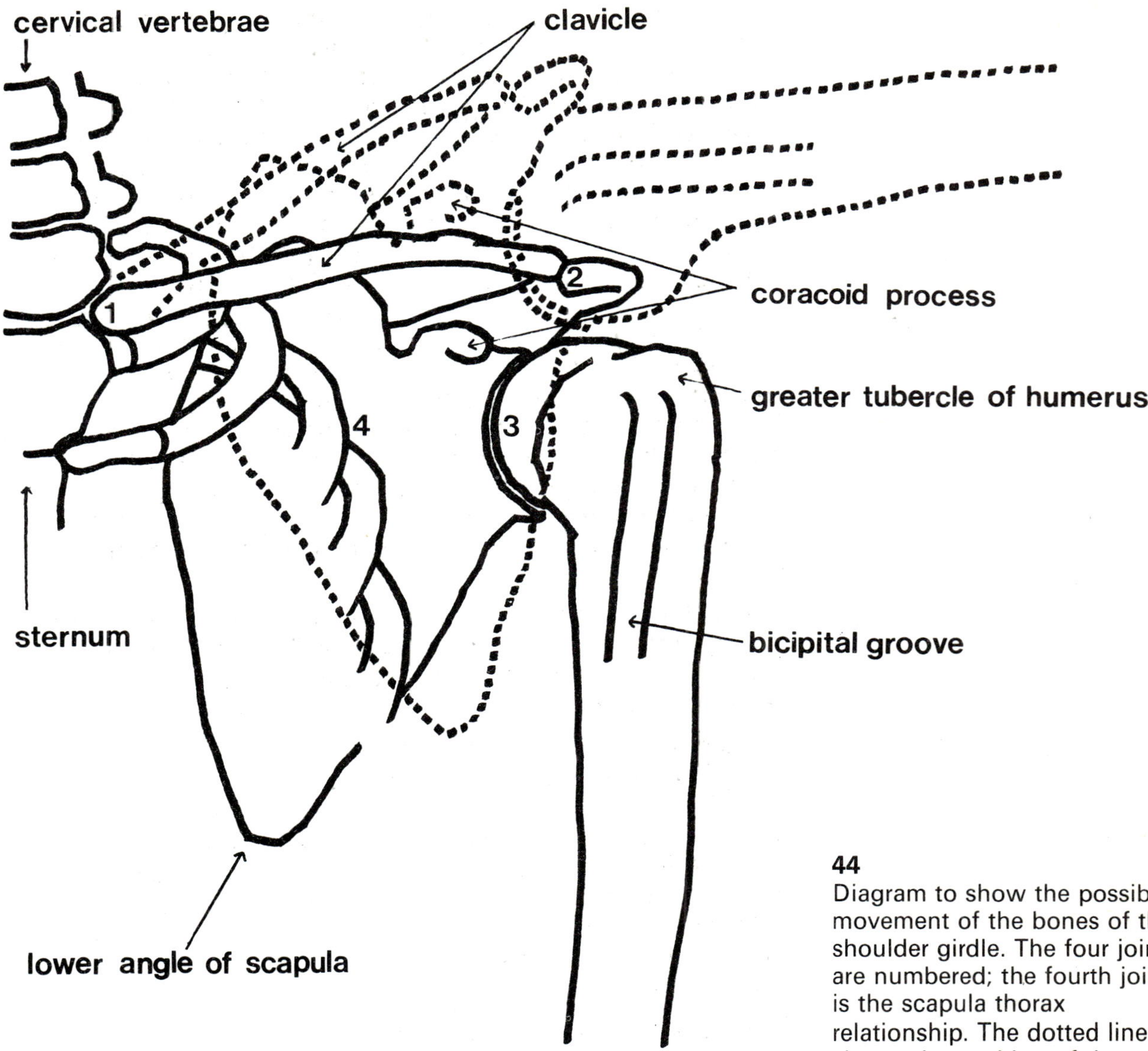

44
Diagram to show the possible movement of the bones of the shoulder girdle. The four joints are numbered; the fourth joint is the scapula thorax relationship. The dotted line shows the position of the bones when the arm is raised away from the side of the body, the scapula having rotated.

contents with laterally-facing bony plates. These bony plates, or hip bones, are joined together at the back by the sacrum and in front at the symphysis of the pubic arch, and they provide firm socket positions into which the head of the femur bone can be joined to form a secure ball and socket joint. (Illustration **43**.)

During walking upper-body weight is transmitted through the pelvis to the legs. The pelvis is built for strength and in itself has a very limited range of movement. Weight from above spreads it out laterally. The pelvic bowl in its external aspects affords mechanically advantageous anchor points for muscles in their action on the thigh bones in walking and running; the pelvis is much broader in man than other animals for this reason. The whole trunk can be assisted to bend forwards when the legs are fixed by the flexor muscles of the thigh pulling from the front surfaces of the femur. Conversely the extensor muscles of the thigh can be used to erect the trunk.

The legs, commencing from the socket in the pelvis, consist of one long upper bone, the femur, and two lower long bones, the tibia and fibular, and then a number of smaller bones which make up the foot. The femoral-tiboral joint is capped by a small separate piece of bone, the patella. The femur bone has a unique and lengthy anatomical neck which facilitates the anterior-posterior movement of the bone by keeping it clear of the pelvic sides, and allows its greater trochanter to be an unobstructed anchorage point, almost in the form of a lever handle, to the muscles which move it. With all tissue removed, the femur direction gives the human skeleton a marked bandy appearance. But it is this inward movement of the femur bones, coming together at the knees, which meets and contains the downward thrust of the mass of the trunk and skull about the centre of gravity. (Illustration **12**.) Strong adductor muscles pulling towards the pubic arch have the almost impossible task of preventing the lower appendages from splaying out. (Illustration **43**.)

The bandy appearance is somewhat more heightened in the female figure than the male, where with rather shorter femur bones and a wider pelvis, the femur pelvic triangle comes slightly nearer an equilateral triangle. The legs are not only capable of anterior and posterior movement, but can also be pulled laterally away from the central body line in abduction and returned in adduction. This, in practice, allows each leg complete cyclic movement when the mass of the body is balanced on the non-proffered leg.

The tibia with the fibula complete the other half of the linear height of the legs, and supply strong and nearly perpendicular terminal posts onto the arch of the feet. The whole leg length can be rotated away from the medial line of the body from the ball and socket joint at the hip. This is a combined movement which with normal muscle action has a 'follow through effect' in the leg and finally effects a lifting action on the arch of the foot. The length of man's legs are necessary to allow for the locomotion movements he wants. This in turn demands powerful muscles; and the well marked ridge on the femur bone, the linea aspera, which is the insertion position of the extensor muscles of the thigh, gives a good indication of the power of the leg muscles by its prominence.

Man's foot is peculiarly his own. He has adopted the plantigrade position and walks on his extended foot. A number of bones of varying size form the foot by a series of arch like forms lying side by side. The arch of the foot, which is maintained by ligaments across the bones, provide an elastic spring board for the whole body mass when walking. (Illustration **45**.)

45
Diagram of the bones of the foot to show their arch-like form. This diagram shows the medial side.

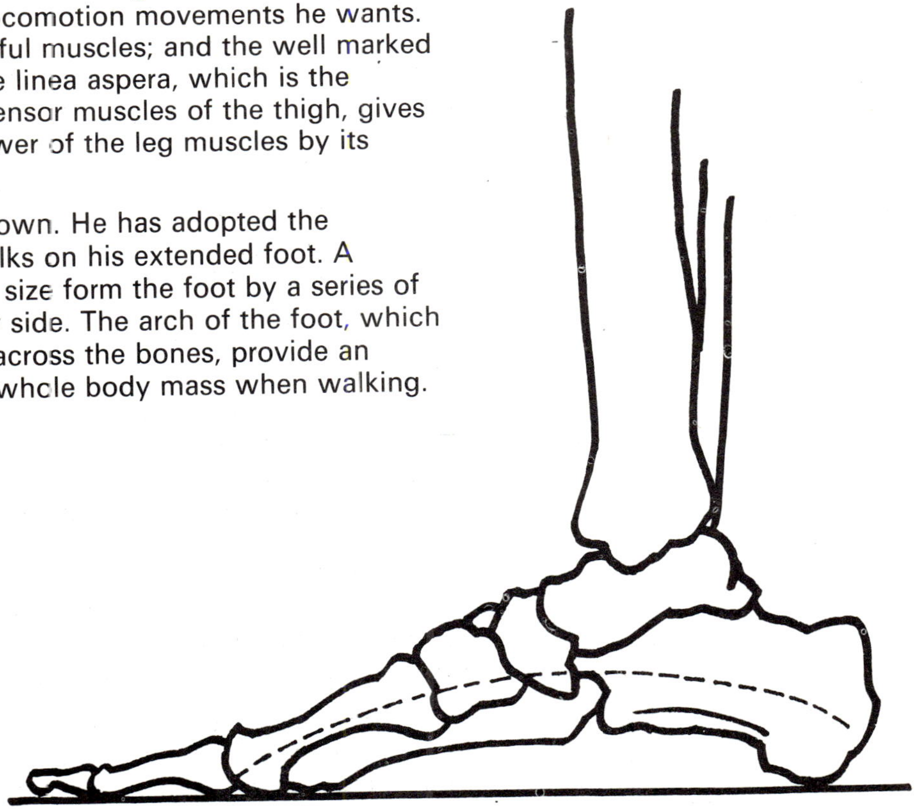

If the motion of walking is considered the conclusion would be reached that there is not just one kind of gait that should be considered normal, because different body physiques give a different appearance to an individual's movement. The head moves up and down in walking, and we are less tall when in motion. The centre of gravity is lower during motion and also shifts sideways from one leg to the other. Body motion requires an inter-play between the loss and recovery of balance.

The arm consists in its upper segment of one long bone, the humerus, in the lower segment two long bones, the ulna and radius, and in the most distal segment of all a number of bones form the hand. The arms, by the nature and relationship of their joints, have a much freer range of movement than the legs. (Illustration **44**.)

The arms and the hands together are motivated grabs which can be extended into an almost undefinable series of positions in relation to the medial line of the body. The arm in total can be passed behind the head, it can also be passed diagonally down behind the trunk; it can also move out in several directions away from and across the front of the trunk. To shorten these distances or to lift or lever the arm can be flexed half way at the elbow. The hand can be rotated into varying positions by the radius bone being passed over the ulna at the wrist end. This action is called pronation; the equal and opposite return movement is called supination. (Illustration **46**.) Varying degrees of pronation and supination can be executed even though the long bones of the arm are presented in a variety of positions to the trunk. Finally the hand itself provides a claw like grip.

This extraordinary range of possible movement stems from the fact that the arm is not fixed to the trunk by one joint, but by four joints. The four joints, progressing from the medial line of the body are: posteriorly, the scapula-thoracic (4); anteriorly, clavicular-sternum (1); then laterally, clavicular-scapula (2); and finally the scapula-humerus (3). (Illustration **44**.) The scapula-humerus joint is a semi ball-and-socket joint. It acts in the manner of a crane-head in all the lifting operations of the human arm, the pivot base of the crane gantry being formed by the scapula-thoracic and the clavicular-sternum joints. The main sideways hoist or shrugging movement with the arms commences at these two joints. The clavicular-scapular joint enables an easement to take place in the bony shoulder girdle, in the position of the crane head, and may offset a possible breakage in what might otherwise be a too rigid system. Altogether the bones of the arm and the hand offer a very flexible series of bony links.

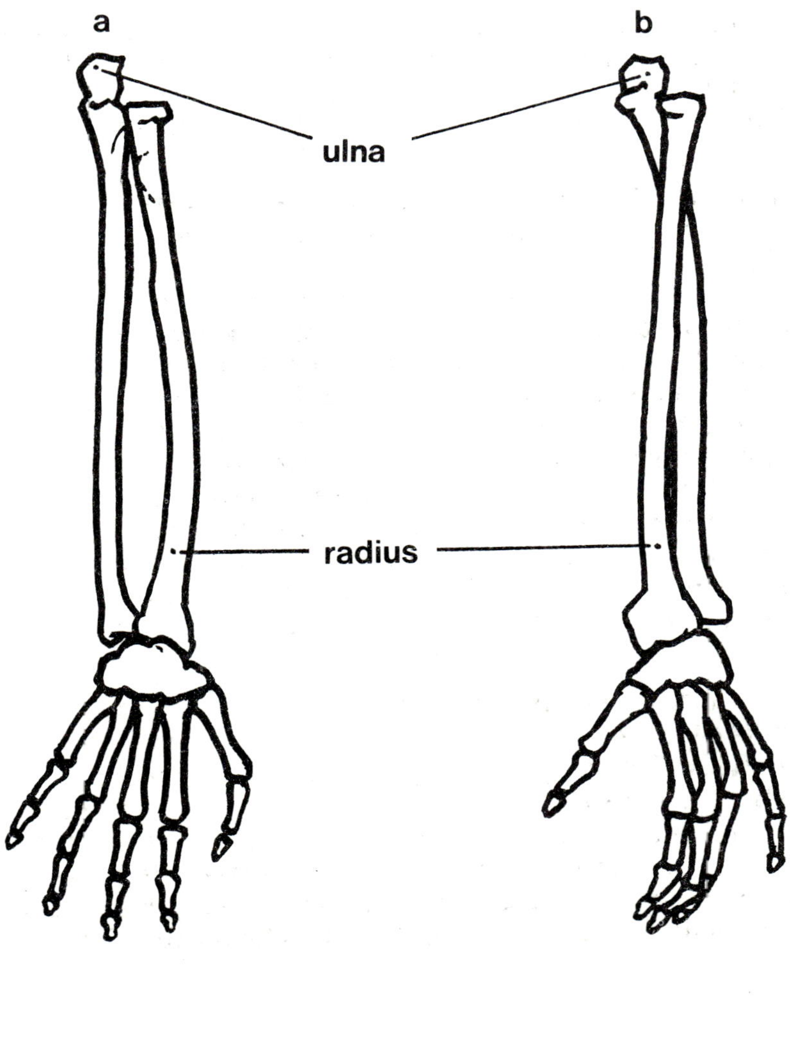

46
Diagram to show the meaning of supination and pronation of the forearm.

a is a front view of the bones of the left forearm and hand. The arm is in a supinated position with the thumb facing laterally away from the body. The radius and ulna lie side by side.

b The arm remains in the same position, but the distal end of the radius bone has turned over the ulna twisting the hand at the wrist so that the thumb faces medially. This is the pronated position of the forearm, and the radius head has been rotated to execute the movement.

The movement of pronation and supination is seemingly increased when the elbow is fully extended because of the addition of some rotation by the humerus bone.

47
A diagram to show the simplified version of a joint. The heavy black lines show the two bony parts in cross section.

The fibrous capsule surrounds the joint completely keeping the bones together. The dotted lines represent the direction of bands of ligaments, which are thickened outer portions of the capsule, the whole binding the bones together with great tension and strength.

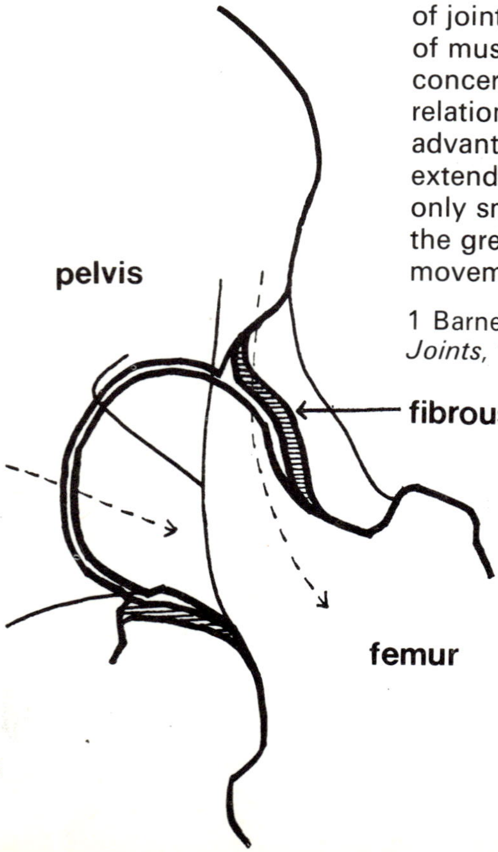

Joints

A joint is the formation, or bringing together, of any two skeletal parts by means of another more pliable structure which allows movement. (Illustration **47**.) The amount of movement at a joint depends largely on the type of surface the two skeletal parts present to one another, and the manner in which they are held together. The type of joint and its position in the skeleton can be related to the demands that are placed upon it. The different types of joints are undoubtedly descendants of earlier varieties which have been modified in the course of time. Contemporary man now has a useful mechanical system though no precise study has ever recorded its many sided capabilities in terms of efficiency. Other animal systems have confined themselves to a more predominant use of the simple ball and socket joint. Human development has expressed a preference for uniaxial joints[1], and it has been further suggested that the selection of this type of joint was evolved because it makes the most economical use of muscular energy. Therefore in any deliberations that are concerned with the human operator and his positioning in relation to controls consideration must be given to the most advantageous use that can be made of the linked flexing extending movement he offers. Two bones in a joint have only small portions of their surfaces in contact at one time, the greatest contact often being found at the limit of a joint movement.

1 Barnett, G. H., Davies, D. V. and MacConaill, M. A., *Synovial Joints*, 1961. Section 2, Chapter 6.

At any position of possible joint movement contact is maintained between the remaining surfaces of bone by smooth cartilage discs. The bone ends are also capped with cartilage, and the joint is securely bound together with ligaments. The cartilage act as a buffer, and because of elasticity of the cartilage discs the fit of the joint is maintained during movement. The cartilage buffer shields the bone ends from wear, and with the maintenance in a healthy person of the self-lubricating system that every joint has, friction is eliminated.

Human joints can either be classified by movement or by the character of the tissues of the union. Classification by movement is the most convenient when studying the type and range of action of an operator's body or limbs. Classification in this manner gives us three types of joint: Immoveable joints, or the joining of two bones which become permanent during the processes of growth, and where no movement can take place. Slightly moveable joints, where a limited movement is allowed. Freely moveable joints, where extensive movement is possible. It is the last two types of joint that merit a designer's attention; and the last type of joint in particular which covers all major executive movements in the limbs. The very free movement of the shoulder joint needs particular consideration by the clothing designer.

The joints of free movement can allow bones to take up varying positions in relation to one another, in all three dimensions. Some joints however only allow movement in one dimension. With all the free joints it is not possible to define precisely their arcs of movement. Distal ends of bones in the appendicular skeleton can point in an infinite number of directions. Every total movement of a limb is an additional sum of the anatomical movement of adduction, abduction, rotation, flexion and extension of the joints of that limb. These movements are explained in illustrations **53**, **54**, **55** and **56**. The range of movement of a limb being increased or decreased by the position of the trunk.

Healthy joints are so manufactured that it is exceedingly difficult in normal usage to overcome their limitations of

48
Diagram to show movements of the joints of the arm. The joints of the leg have the same series of movements; multiaxial at the thigh, biaxial at the knee and uniaxial in the foot. The great difference between the hand and the foot are that the phalanges can be opposed across the palm of the hand, whereas the phalanges of the foot make a series of parallel arches.

movement and damage them. Joints can resist a sudden force as well as a continuous one.

Arcs of movement in all three dimensions depend on bone construction, associated muscles and ligaments. Joints in the appendicular skeleton can become stationary by the checking action of the attached muscles and ligaments. Conversely, some rotation can take place at a number of limb joints. Alternatively, rotation of the distal end of a limb can be increased by rotation at the proximal end of limb at the trunk. For example the action of the buttock muscles rotates the femur bone, which movement is transmitted right through to the foot, making it conduct a lateral rotating

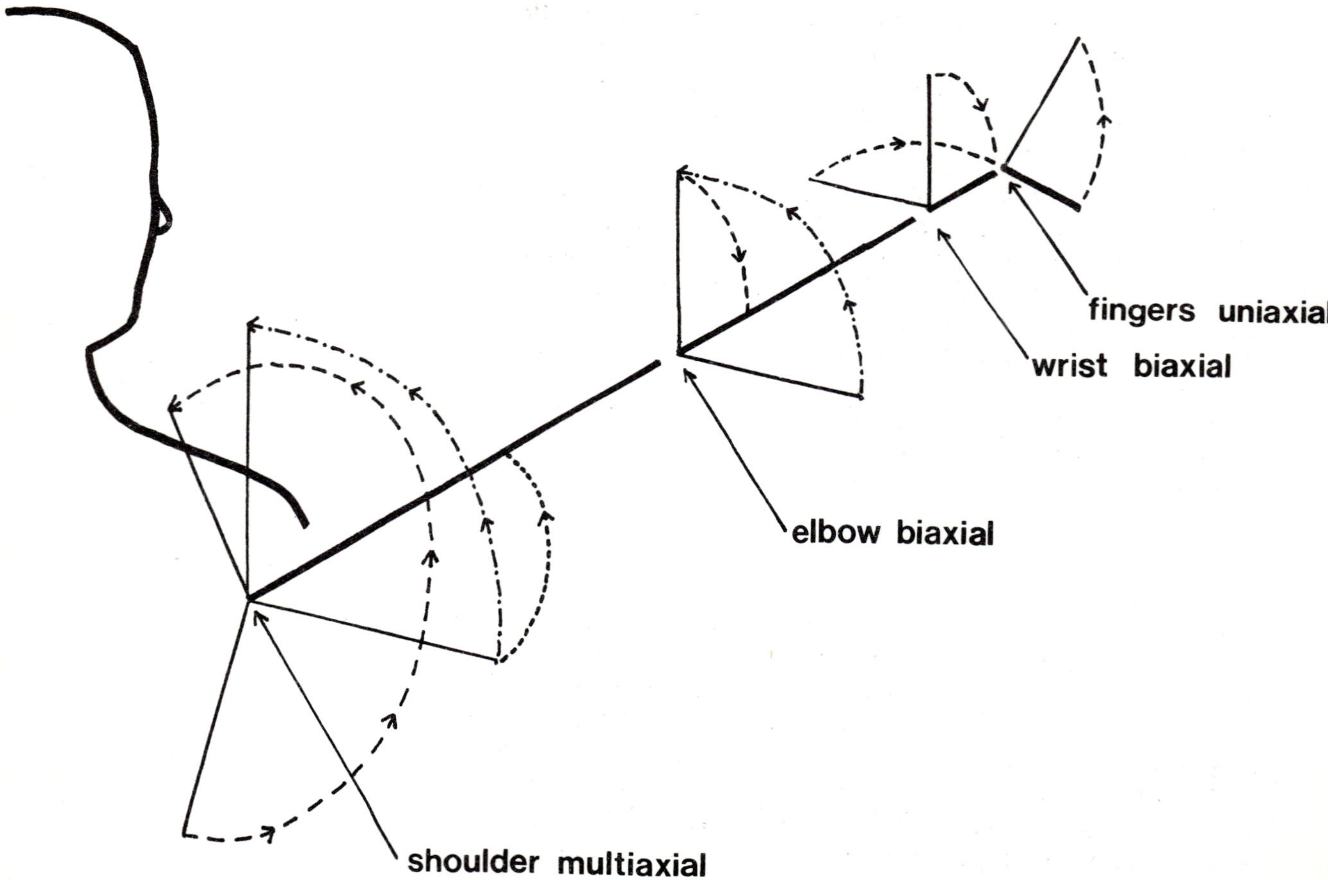

movement. Alternatively, when the tibia bone and femur bone are at right angles, the tibia bone is able to rotate to some degree with the foot laterally.

The arm possesses much free movement (illustration **48**). In certain arm positions its swing takes it through three dimensions of movement simultaneously, in other positions different dimensions of movement have to be achieved independently. In other combined movements from the shoulder the rotation of the humerus goes unnoticed. The arm fully flexed at the shoulder offers the operator a more deliberate torque at the wrist. It can well be used in the standing position where reach may be extended by body movement. When the arm is raised at the shoulder by flexion shoulder and wrist action combine well together to facilitate deliberate hand use on large scale. Although these predominately shoulder movements will require practice on the part of an operator.

With the elbow in a partially flexed position an operator can conduct slight pronating and supinating movements for calibrated dial adjustment with near viewing. Because in this position the elbow and wrist joint movements may be combined to offer the widest range of movements for the hand. Most joint movements are of a conscious nature in the conception of an action, a vital proportion are actionable by conditioned reflexes and establish everyday positions the body requires. Quick movements to escape damage to vital parts come into the same category.

Muscles

Muscles work by contraction. They consist of bundles of fibre tissues. A muscle fibre can shorten its length by at least one third in contraction. (Illustration **49**.) Contraction brings the two ends of a muscle nearer together. Therefore it brings the two bones to which it is fastened nearer together. Muscle as contractors always pull. Even when, for example, we push down with our forearm it is because of the pulling action of the triceps in the upper arm. The energy of a muscle is

49
Diagram to illustrate the simple action of a muscle. When a muscle contracts it shortens its length. With its two ends attached to different bones by tendons upon contraction one or the other of the bones must move by pivoting at a joint.

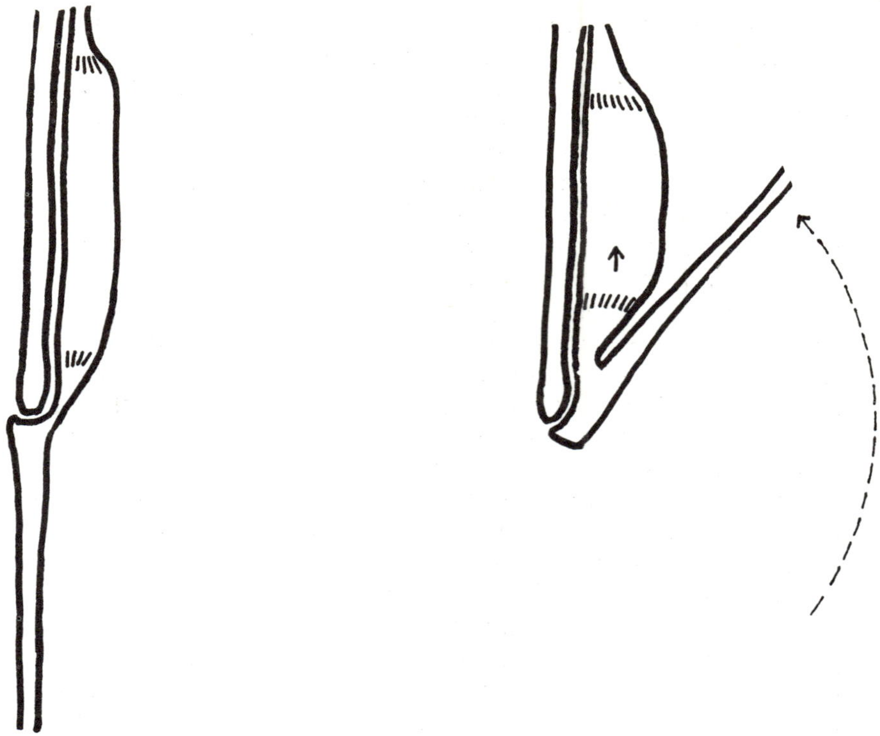

expressed in contraction against weight or resistance and is the result of bio-chemical processes. These processes are triggered off by impulses along the motor neuron of the nervous system from the brain. The actual bio-chemical nature of the impulse along the neuron, which result in the contraction of muscle tissue, are not yet fully understood. Any movement requires the use of a number of muscles and their actions are co-ordinated by the motor nervous system.

A muscle is divided into two kinds of tissue: A contractile portion between two tendons which are non-contractile. The two tendinous portions are generally designated the origin and insertion of the muscle. The origin may be thought of as the proximal portion of the muscle, from which in terms of pulling, the action of the muscle may have said to have originated. Muscles act as prime movers when they initiate an action: as antagonists when they act against prime movers to control the rate of action; as partial antagonists when they assist the prime mover to make one definite action; and finally as fixators or stabilisers when they hold parts of the skeletal frame firmly together whilst an action takes place elsewhere. Muscles rarely function alone, and most are composite in character and can undertake more than one function at a time. It is possible for one portion of a muscle to be active in a major role, whilst another portion is playing a minor role, and yet a third portion showing no activity at all. So that muscle action is often the co-operative action of a number of muscles, or some of their parts. Thus compositely the actions of separate muscles shade together to effect varieties of movement and to give them clarity and precision. Composite muscle action is often difficult to assess in terms of what number of muscles or parts of muscles are responsible for particular movements. It is quite possible to use an alternative sequence of muscle actions to effect the same kind of movement. This has been demonstrated by the achievement of subjects who have temporarily or permanently suffered partial limb damage. The composite actions of muscles suggest that when consideration is being given to the situation of a human operator, provision should be made for space to allow a shift of position so that fresh muscle relationships can be tried and the operator enabled to carry on. Shifts of position offset fatigue.

Antagonists pull against prime movers even in so-called rest. Some muscle energy is therefore always being expanded. This quiet pull of muscle against muscle is called tonus, and adds to the general feeling of well-being in a healthy person.

Cramp, or the locking of muscle contraction, can come about by over exposure to low temperatures, over-exertion or a lack of common salt.

Part of a muscle's duty in composite action is to exchange roles to prevent damage to a joint. Thus the flexor muscle of the arm, the bicep, becomes an antagonist to the tricep muscle a fraction of a second before full extention of the arm at the elbow is completed. The bicep pulls momentarily at the termination of the movement. This pull is a result of a co-ordinated reflex action by the nervous system. A further reflex action for the contraction of the antagonist muscle is initiated from the ligaments and other tissues binding the joint itself. In normal movement the reflex action that originates from joint tissue becomes the second means of deaccelerating the action of a joint.

Reduction of the use of joints in a limb may lead to economy of an operator's energy. For instance the use of the whole arm-reach without movement at the elbow or shoulder is much less tiring for short periods than the continuous use of all the intermediate joints; because the olecrannon process of the ulna fits into a corresponding fossa of the humerus when the elbow is extended, and the manner in which the two bones can rest together allows an amount of relaxation for the surrounding muscles. However in the consideration of the use of limbs by an operator a careful evaluation of their task is necessary, and the conservation of energy may be seen as secondary to the effectiveness of their operations.

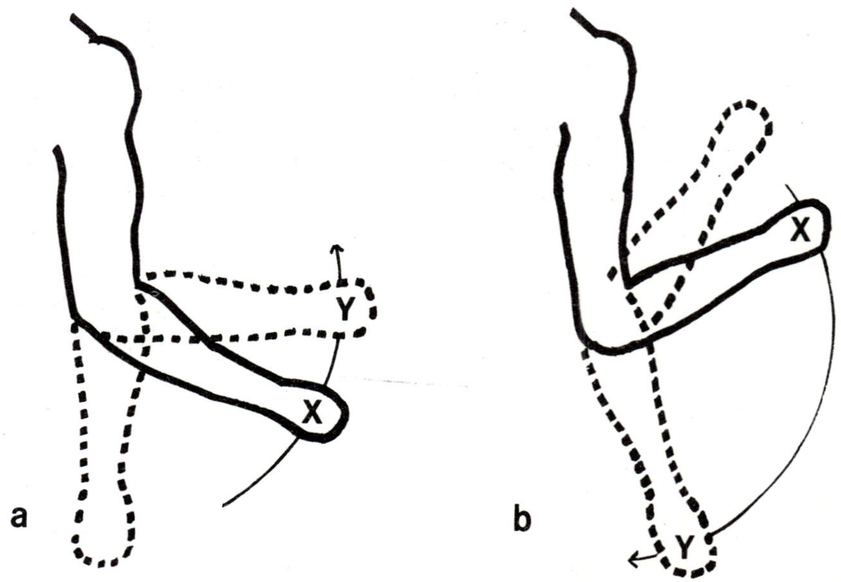

50
a illustrates the arm in a midway position towards full flexion. Power is greater at position X to overcome resistance than at Y where it would decrease rapidly.

b illustrates the arm in a midway position towards full extension. Power is greater at position X to overcome resistance and it decreases to Y and beyond.

Considerations towards the use of Human Limbs and Joints in Operator Functions

Forces should only be moved by limbs whose joints are in a 'midway' or intermediate position (illustration **50**). This is because the nervous reflex action of joint ligaments, which acts as a second safeguard to avoid joint damage only takes place when the joint is within a few degrees of the completion of its movement. The 'midway' position for commencement of exertion allows the antagonist reflex action as well as that of the ligaments to take place in the optimum period of time. Thus the two reflex actions are able to prevent joint tissue injury. The resistance of the force may also be too great for the operator to overcome. Attempts to overcome resistance or force from the 'midway' position allows the operator to naturally test their capacities. Trial against resistance when the joint is near the end of its movement not only causes anthropathic effects, but positive damage.

There are further reasons why it is important that forces should be countered by limbs whose joints are in a midway position. The joints related to physical movement, together with their attendant musculature and bony extenuations, can be seen by their arrangement to be acting as first or third class levers. (Illustration **52**.) There is continuing debate, about the presence of second-class levers. We may assume that our place in the animal kingdom has depended amongst other things on our evolution showing a preference for the first- and third-class levers. Be that as it may, the fact that we do use levers is relevant to the problem of the consideration

51
An example of a human lever. The forearm working as a third-class lever.

F The fulcrum is the distal end of the humerus bone.

L The load is in the palm of the hand.

E The effort is the contraction of the biceps muscle which is overcoming the resistance of the load and causing movement upwards shown by the dotted line.

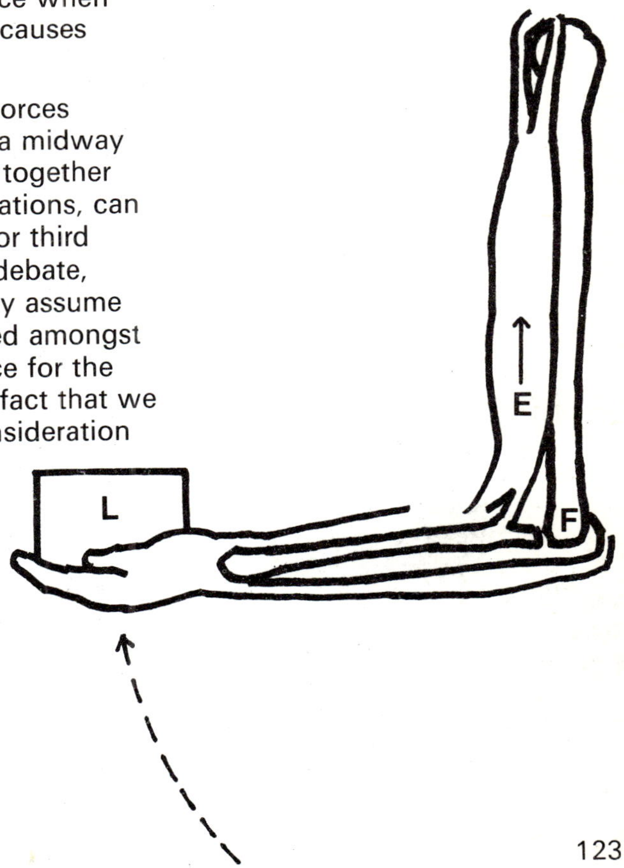

of the most efficient way of using our physical powers. The power of a human lever in exertion depends on the power or the effort the muscles can apply to the bony lever to obtain equilibrium with, or to overcome, the resistance being applied to it, and the power that a muscle can offer is greater at the beginning phase of contraction than at the termination.

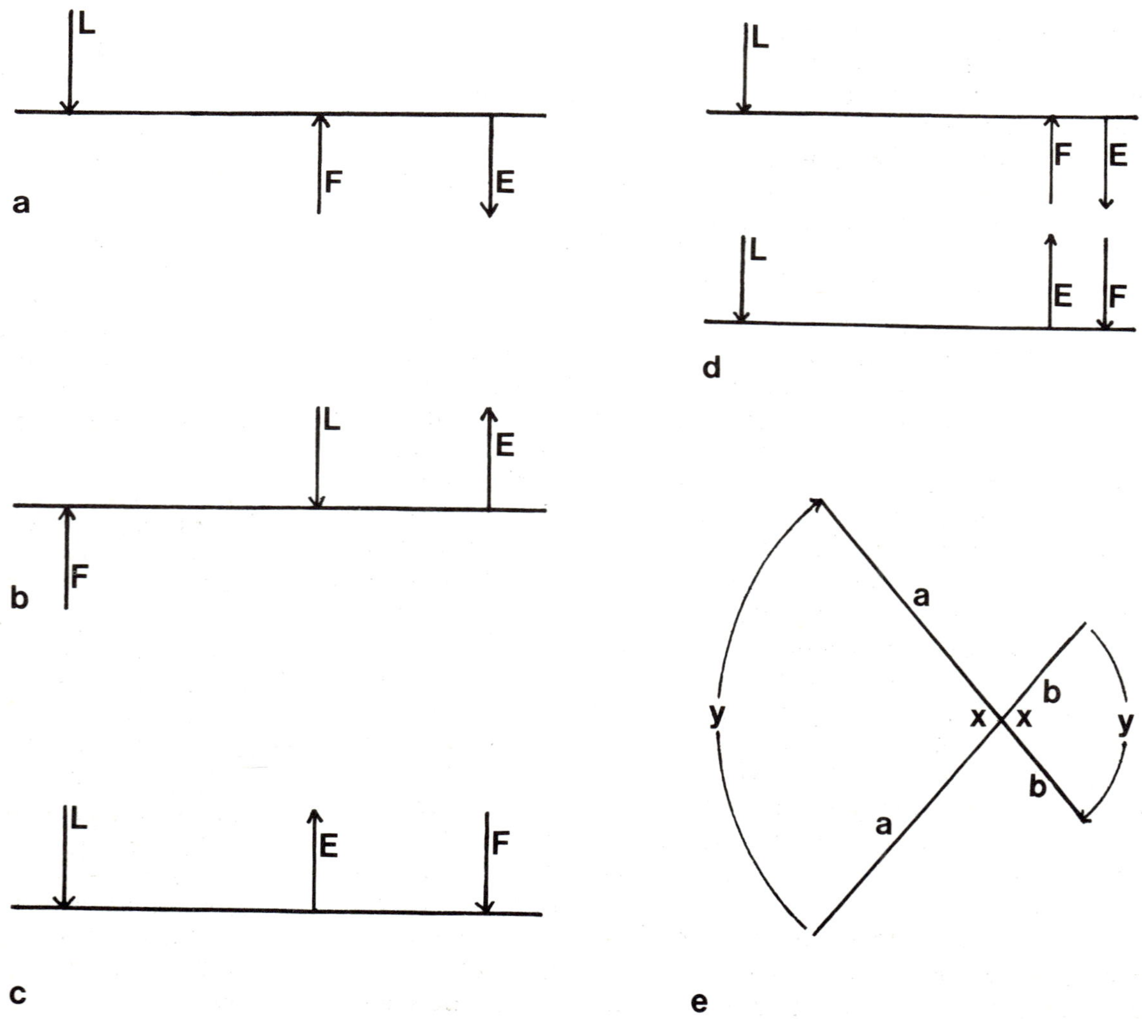

Midway positions, where muscles are only partially contracted therefore give larger forces, or the possibility of more effort. (Illustration **50**.) The 'midway' position also allows the prime mover muscles which give power to a movement to operate fully; at the termination of joint movements only the stabilising muscles which hold the bones of a joint together are functioning, and as their name implies, they cannot exert force away from a joint.

Third- and first-class lever lengths can easily be increased effectively; that is to say the points of the fulcrum and of the application of power can be kept together, whilst the application of the load or resistance can be moved further away. (Illustration **52d**.) As the length of the lever increases so the velocity at the load end can be increased. (Illustration **52e**.) This is precisely what has taken place in the evolutionary construction of man's limbs, and it is therefore possible for us to move optimum loads with speed.

With limb movement the correct speed for the action is important to the human operator. Speeds of operation which are too quick or too slow cause fatigue; the fatigue being generally caused by unnecessary complex muscle control due to unnatural movement and positioning. Complex muscle control of joint movements is wasteful and inefficient, when it only produces a small effect that could have been produced by simpler means. The operator being placed too near or too far away from controls are both positions where muscular and nervous energy would be wasted, for in both postures the limbs would be over extended or uncomfortably flexed, and their use in this position would make unnatural demands on an operator. For example, too great a flexion of the knee will restrict the smooth operation of ankle or foot movements. Too great a flexion of the elbow will restrict the smooth operation of wrist and hand movements. When the leg is partially extended foot movement on pedals are under far better control. When the elbow is partially extended, pronation and supination at the wrist is smoother, and the hand is able to offer a more deliberate form of control. For very delicate control movements the forearm may be retained near the operator's body and the hand held slightly below elbow level.

52

a, **b** and **c** show the three orders of levers:

L = The load or resistance to be moved
F = The fulcrum
E = The point of application of the effort or force

a is first-class, **b** second-class and **c** is third-class.

d illustrates how first-class and third-class levers can be effectively increased in length by the point of effort and the fulcrum being brought together.

e illustrates how the lengthening of a lever can increase the velocity at the load end.

Levers a and b have the same angular velocity at x. a however is longer than b, and at the load end y a has a much greater linear velocity than b.

An example calculated from human movement shows that the hand swung by the extended arm from the shoulder moves nearly three times faster than the hand swung by the forearm from the hip.

53

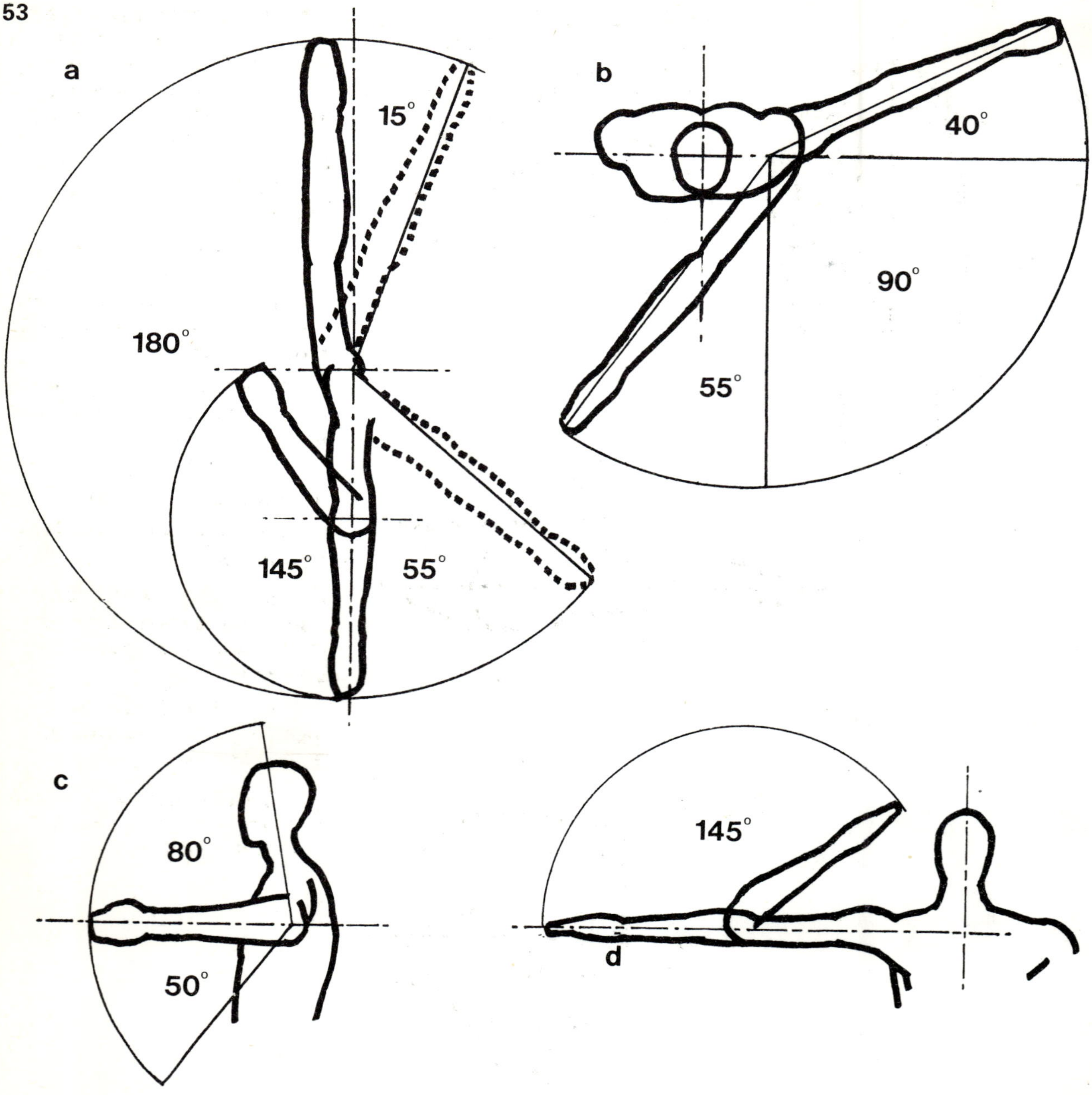

54

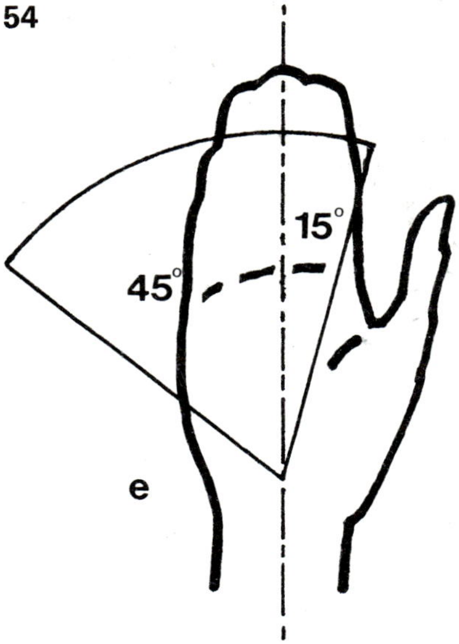

e

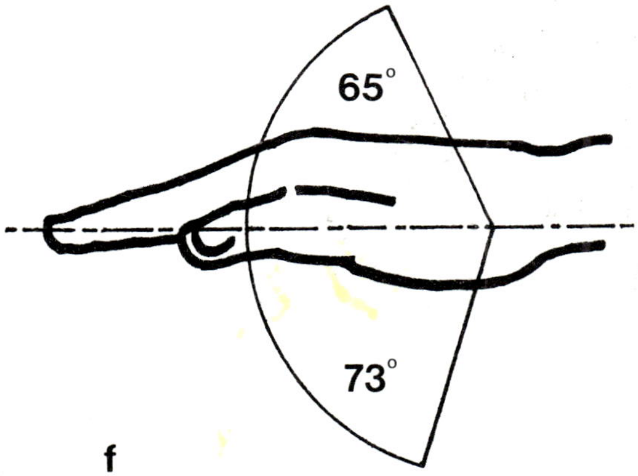

f

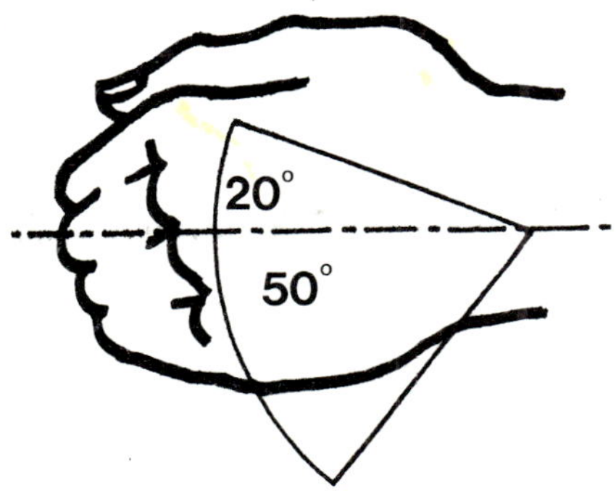

g

53, **54**, **55** and **56**
Diagrams to show the capacity for angular movement of the limbs or limb segments. The angles stated have been estimated from a variety of surveys, and in most cases there is not just one angle of movement of which everyone would be capable. As with other forms of measurements there are gradations of values in terms of percentiles. For critical operations retesting would have to be carried out with the population of users concerned. The angles stated should cover the contingencies of everyday operations.

a Shoulder flexion from the horizontal to behind the head. Shoulder extension to behind the centre line of the body. Elbow flexion.

b Shoulder hyperextension behind the body, and shoulder adduction bringing the arm across the body.

c Rotation of the elbow.

d Elbow flexion towards the medial body line.

e Hand palm facing. Wrist adduction towards the body medial line, and wrist abduction.

f Wrist dorsiflexion above the centre line, and palmar flexion. Forearm pronated.

g Wrist flexion and extension with forearm supinated in the perpendicular plane.

h Hip flexion above horizontal line, extension and hyperextension behind the line of gravity.

i Hip rotation. Lateral and medial.

j Hip swing. Lateral and medial.

k Knee flexion.

l Hip adduction across medial body line, and abduction.

m Ankle dorsal flexion above the centre line, and plantar flexion.

55

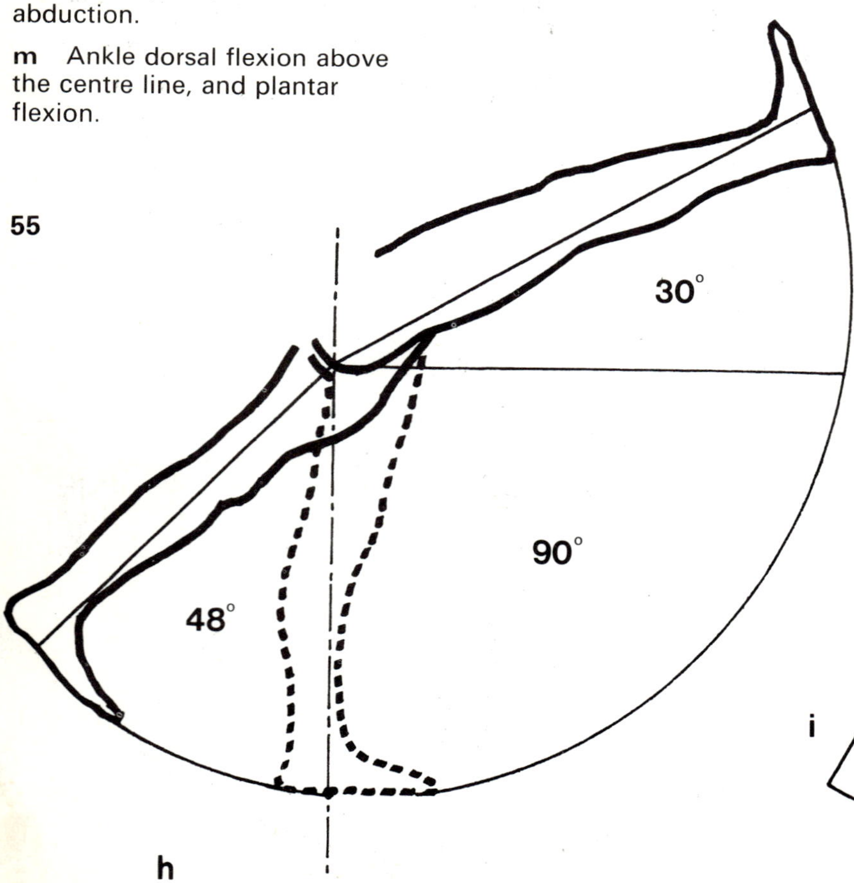

h

i

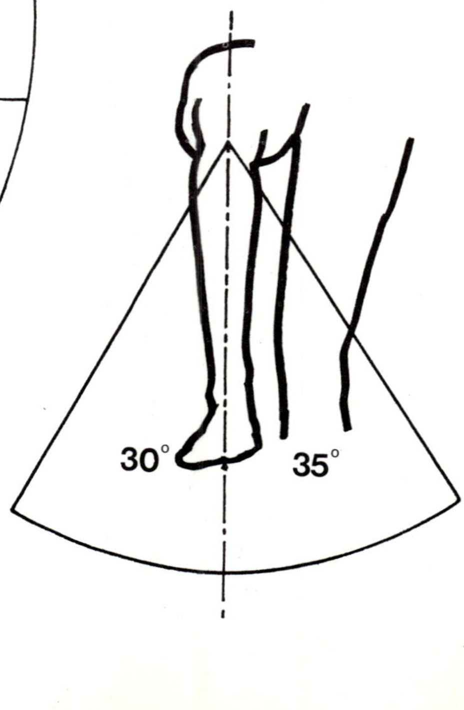

exercise

56

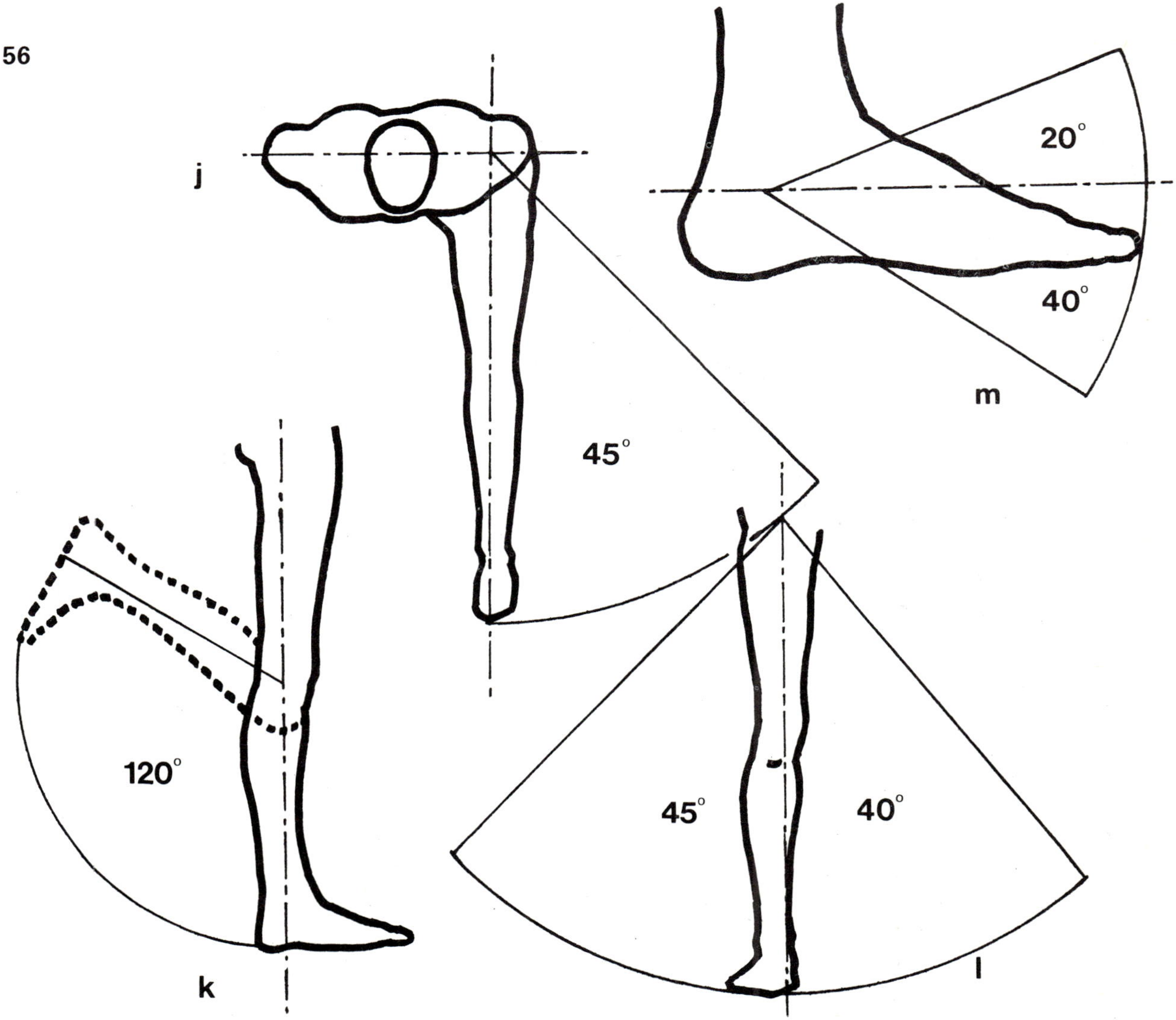

Body and Limb Dimensions related to enclosed and Working Environments

Movements of force, or movements with levers or wheels are less awkward and quicker if they can be arranged to move towards the medial line of the body or downwards following the force of gravity. It therefore follows that anti-clockwise movements are easy and efficient, particularly when arranged to take place in a vertical plane. (Illustration **57**.)

Levers

Levers can be arranged to give a wide range of movement; the greatest range of movement being obtained when they pass across in front of the body, rather than towards or away from it. Greater force can be applied by the second type of lever movement. Lever forces are not suitable for fine adjustment, but they can give positive on/off switching or positive multiple switching providing the range of intermediate contacts or position are neither too great in number or too close together.

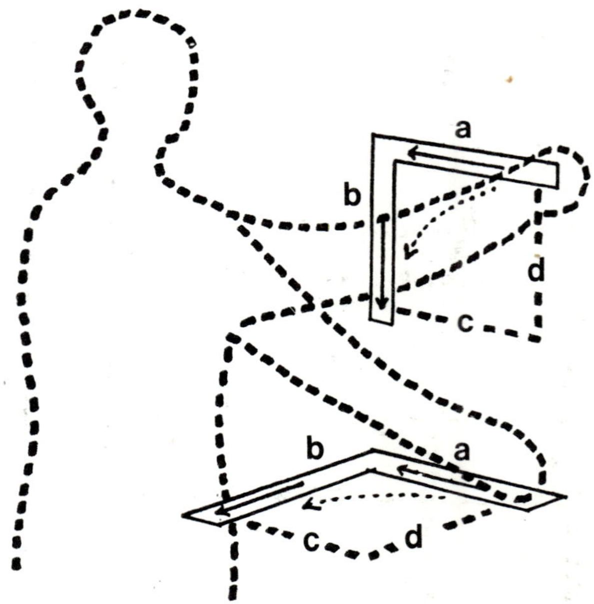

57

Control movements made inwards and towards the body can be faster and superior to movements made outwards and away from the body.

They can be equally effective in the perpendicular plane.

The control movements can either be angular, or as shown by the dotted arrows, follow a curved path.

Paths a and b in both cases would be actionable movements; c and d return control paths.

Large forces can be applied by a lever particularly when an operator is standing, and greater again if the weight of the entire body is allowed to contribute. For the reasons that have been discussed elsewhere this is unwise and should not be allowed to take place, and levers should be so sprung that only minimal or medium effort is needed for their movement. A travel distance of 4″ (102 mm) can be quite satisfactory for a short lever.

Lever Loadings for Operator in sitting position. The lever moving up and down in the Median Line of the body.
A good arm-position for this type of operation is when the upper arm is approximately 30° to the centre line through the trunk, loading amounts decrease above and below this angle. 5th percentile values for these two movements are:

Pull down 20 lbs (9·07 kgs); Push up 15 lbs (6·8 kgs).
(Illustration **59b**.)

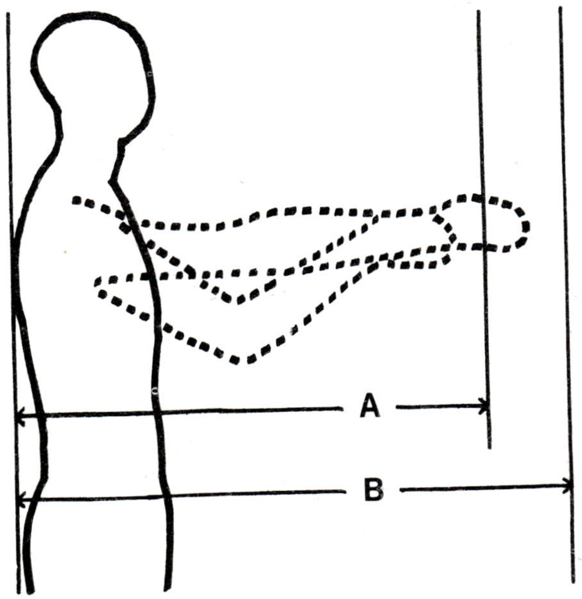

58
A favourable position for maximum pull and push movements.

A Distance towards which maximum pushing could take place is 29″. (737 mm).

B Distance from which maximum pulling could take place 34″ (864 mm).

59
a Illustrates sitting position for operating a lever towards and away from the body. Difficult arm position is dotted. More favourable position for arm is when upper arm makes an angle of 60° (x on diagram) with centre line of trunk.

b Illustrates sitting position for operating a lever up and down in the median line of the body. Angle (x on diagram) of 30° of upper arm to the body is the most favourable position.

Lever loadings for operators. The lever moving towards and away from the body in the Median Line, with the lever in a more favourable position from the body.
Values given for left arm. This is normally the weaker arm so that 5th percentile values provide maximums within the capacity of all for both arms. (Illustration **59a**.)

Maximum pull only when arm is fully extended.
Maximum pull 5th percentile 50 lbs (22·7 kgs).
50th percentile 110 lbs (49·9 kgs).
Maximum push when arm is not fully extended.
Maximum push 5th percentile 40 lbs (18·14 kgs).
50th percentile 120 lbs (54·43 kgs).

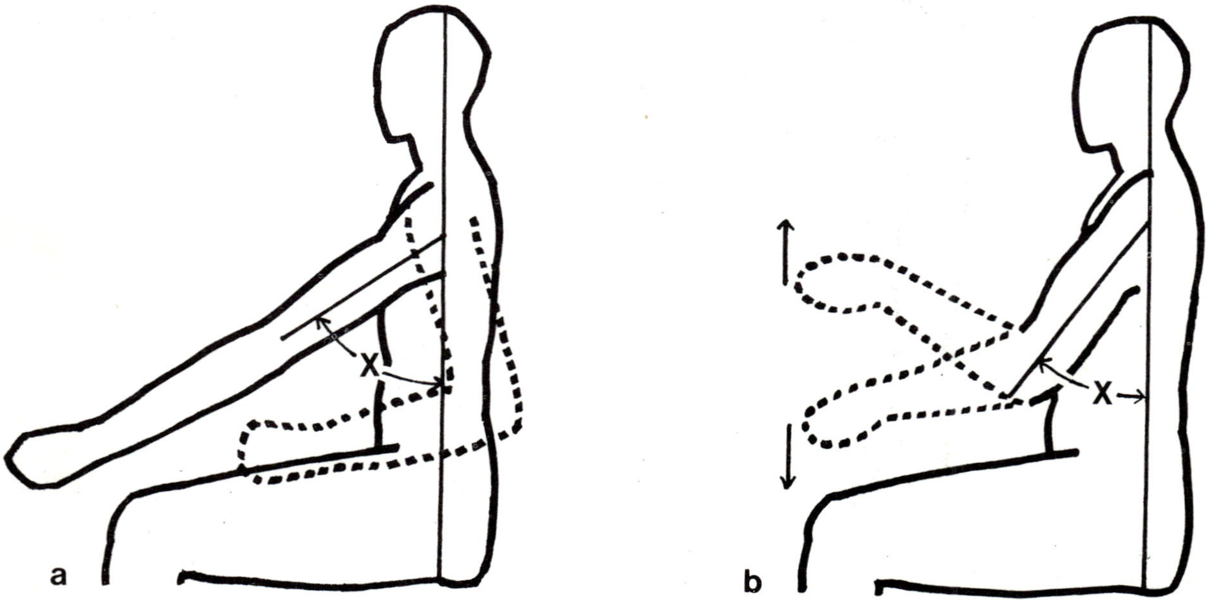

Lever loadings for operator in sitting position. The lever moving towards and away from the body in the Median Line.
The Maximum load for the most difficult position for the left

arm for the 5th percentile operator is 20 lbs (9·07 kgs). More favourable positions for the left arm for the 5th percentile operator allow loadings of 30 lbs (13·6 kgs).

For push-pull levers 20 lbs (9·07 kgs) can be taken as a very safe loading; 30 lbs (13·6 kgs) therefore, can be taken as safe loading figure and it is the one recommended.

Lever loadings for operator in sitting position. The lever moving across in front of the body laterally.
5th percentile values for this movement is pulling in 15 lbs (6·8 kgs); pushing out 15 lbs (6·8 kgs). (Illustration **60**.)

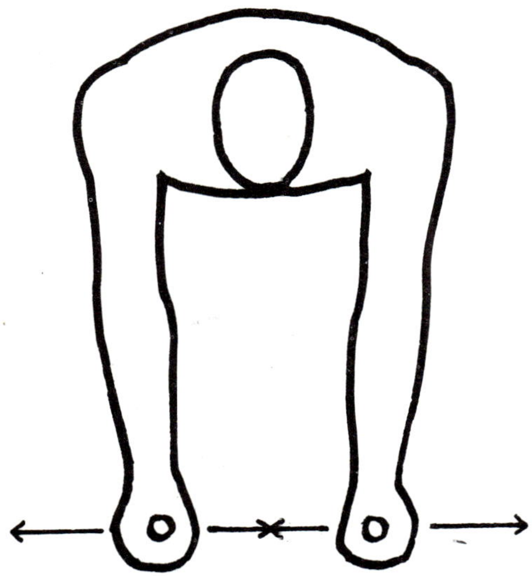

60
Lateral Lever movements.

Pedals

Total leg strength is next to total postural strength in power, and the force exerted by their extension is greater than that obtained by flexion.

A large amount of force can be transmitted with leg extension by a seated operator using a back rest, although this cannot be sustained for any period of time. Although offering great power, leg and foot movements are not accurate. They are good for start/stop, on/off switching; steady pressure (although this may require footguards to maintain the foot on the pedal); and steady repetitive motion.

Pedals can either be operated by the whole leg and body (with or without back rest) or the foot pivoting at the ankle. (Illustrations **61**, **62** and **63**.) Foot pedals exert less force, but they do offer more accuracy because conscious control of the nerves and muscles acting on the feet can be co-ordinated to the degree of unison of action that will produce graduated pressures. But in general the action of the legs and feet in the majority can in no way approach the refinement of action and tactile awareness of the arms and hands.

Pedals cannot be operated satisfactorily or comfortably from a standing position. Both feet are normally used in standing to preserve balance; if a floor pedal is used it should be at ground level and have a very short travel distance, so that the operator's two feet remain as near as possible to the horizontal.

Without undue effort any thigh and lower leg angle between 90° and 170° can offer a force of 30 lbs (13·6 kgs). (Illustration **63**.)

Much greater forces can be exerted by foot pedals, the effort required being minimised by the operational leg being 'sighted' in a particular angular relationship. Typically this is an angle of 135° between the thighs and the lower leg. (Illustration **61**.) With this angle or slightly greater forces of 300/400 lbs (136 kgs/181·43 kgs) can be offered. Thrusts with the leg to achieve these forces or greater are very quick and the counter pressure of a back rest is needed. The angle of 135° is critical as a minimum; there are degrees of tolerance towards a larger angle but the limiting factor comes with most figure types before the leg has reached two-thirds of full extension. This is because of the lower legs action as a third class lever in extension.

With effort larger forces can be exerted in this position, but only for fractions of a minute.

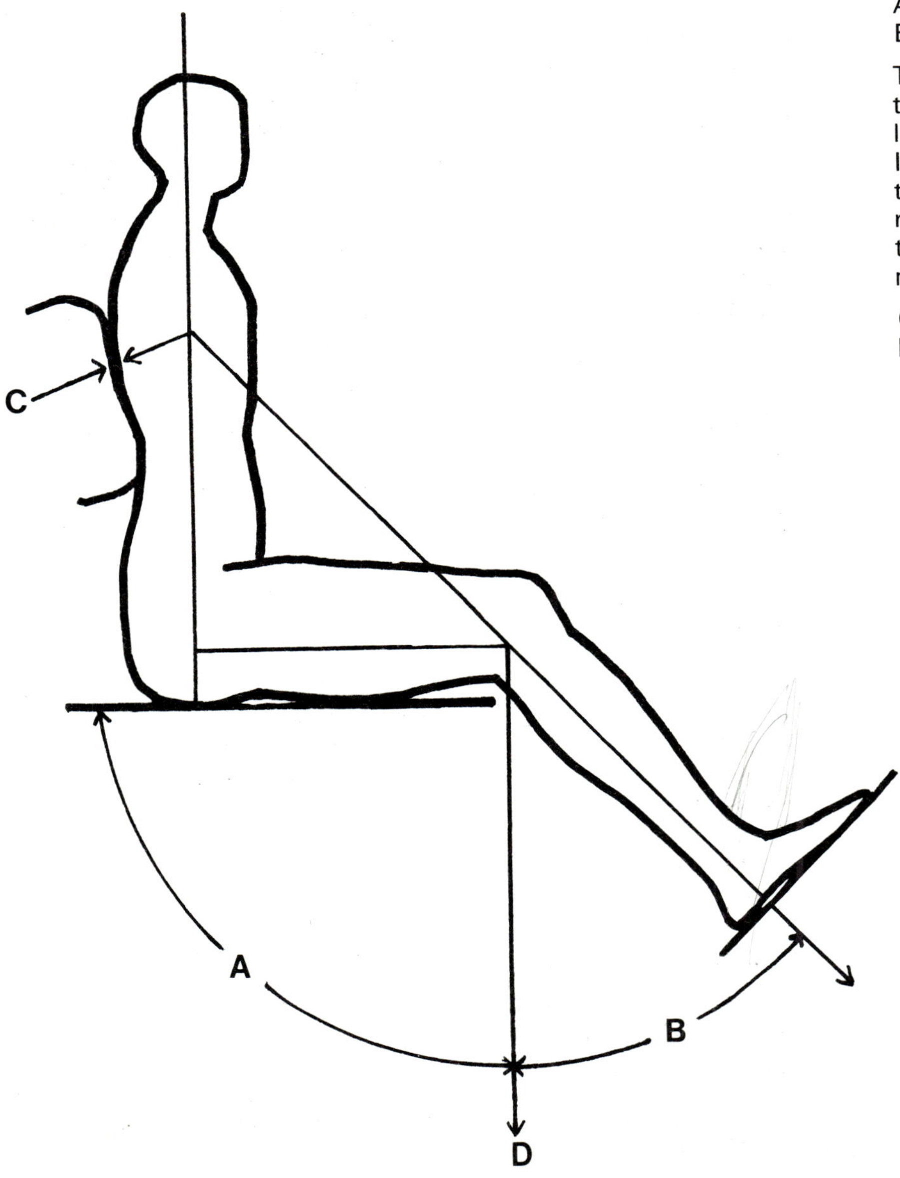

61
Leg position to achieve large forces of thrust.
A = angle of 90°
B = angle of 45°

This position allows maximum thrust. Thrust at D would be less than 100 lbs (45·4 kgs). If any adjustment were made to the seat back and seat pan relationship, the 135° angle of the leg would still have to be maintained.

C Counter pressure supplied by stable back rest.

62
Foot pedals
Foot should make an angle of 90° to 100° with lower leg. If the angle of the pedal surface to the ground is 15° or more a heel stop should be provided.

An optimum range of pressures possible for continuous work would be from 3 to 12 lbs (1·4 to 5·44 kgs). A pressure of 20 lbs (9·07 kgs) or more should be applied by the whole leg.

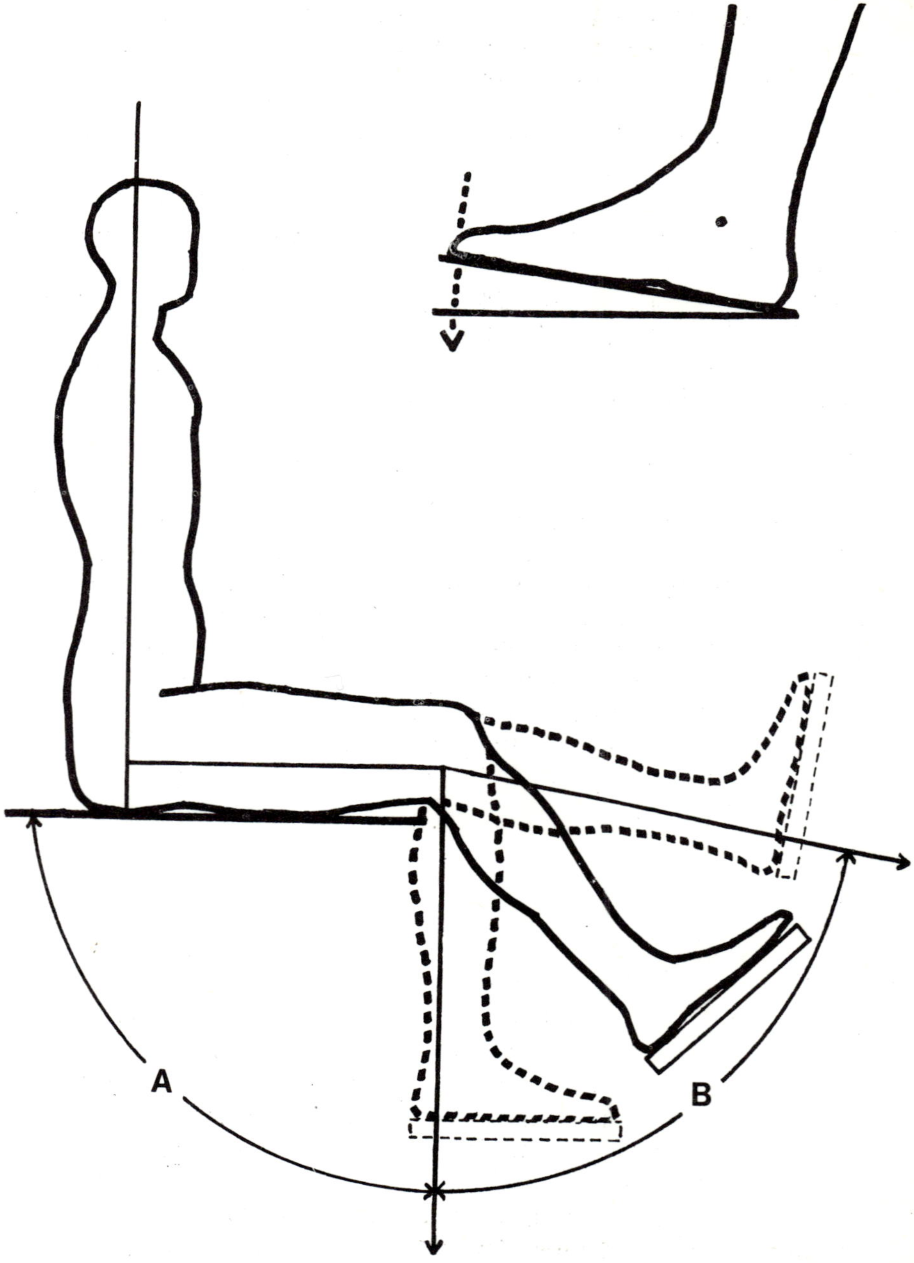

63
Leg positions for light thrusts of 30 lbs (13·6 kgs).
A = angle of 90°
B = angles between 90° and 170°

The pedals are at right angles to the lower leg.

Adjustable or sliding seating should be provided for pedalling positions.

If possible pedals should be spaced 12″/14″ (305 mm/356 mm) apart, their centre being in line with the median line of the body.

90°/100° should be maintained between foot and operated surface as with foot pedals.

Various angles of leg presentation and body stance can therefore present a gradation of forces from 30 lbs (13·6 kgs) to 300 lbs (136 kgs) or more.

Percentile values for feet dimensions in inches and millimetres

	5th		50th		95th	
	in	mm	in	mm	in	mm
Foot length Male	9·8	249	10·5	267	11·3	286·5
Foot length Female	8·8	223	9·5	241	10·2	259
Foot breadth Male	3·6	91	3·9	98·5	4·3	109·5
Foot breadth Female	3·3	83·5	3·6	91	3·9	98·5

Cranks A handle mounted on a turning wheel
Cranks up to about 4″ (102 mm) in diameter give faster speeds. After this size the speed decreases as the size of the wheel increases. They offer a good degree of accuracy and particularly the larger wheels can be used where the control of adjustments can be arranged to follow continuously and full circle. They can be used to move loads of 30 lbs (13·60 kgs) or more. The shaft of the wheel can either face the operator or be set parallel to the way he is facing. (Illustration **65b**.)

Handwheels
Used by both hands and give good operator stability. They offer reasonable powers of adjustment through a turn of 90° or less. They are best when placed facing an operator. When mounted with the wheel in a perpendicular position they can exert greater force when turned in an anti-clockwise direction. They can be used to move heavy loads in the same way that a lever can. However, this could involve the use of the operator's body weight, and as with lever operations, should be avoided if at all possible. They are best used with moderate loads of 40 lbs (18·14 kgs) to 50 lbs (22·7 kgs).

Their diameter should be in the region of 12″ (305 mm) to 14″ (356 mm). (Illustration **65a**.)

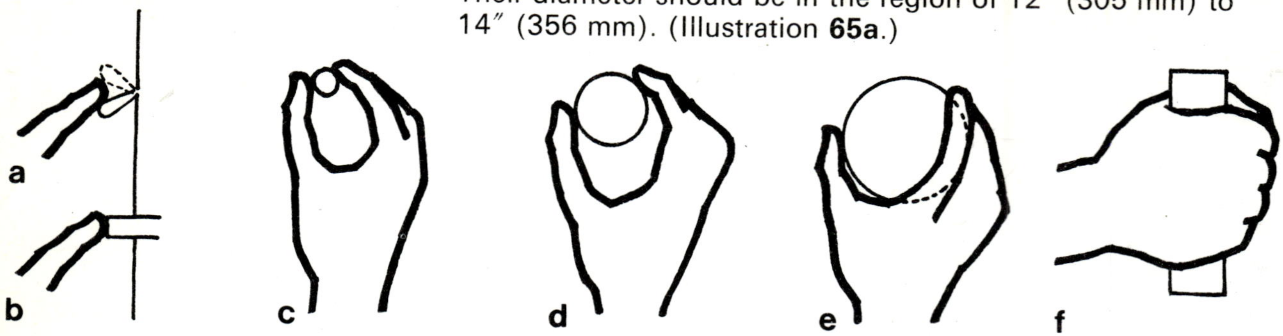

64
Different finger and hand grips.

Knobs

Knobs are used for a finger or precision grip for making fine adjustments. Small sizes may be only lightly loaded. $1\frac{1}{2}$″/2″ (38 mm/51 mm) is a good size and offers smooth operation and accurate adjustment. A 2″ (51 mm) diameter knob can have loadings from 2 to 8 lbs (0·9 kgs to 3·63 kgs). (Illustration **64d**.) A 3″ (76 mm) diameter knob or more requires operation by wrist movement. (Illustration **64e**.) Knobs of this size may be more heavily loaded, and between 3″ (76 mm) and 5″ (127 mm) can take 15 to 20 lbs (6·8 kgs/ 9·07 kgs) loading. Knobs greater than 5″ (127 mm) become impossible to hold easily to produce a torque by operators with 5th percentile hand widths. Greater torque values must be produced by handwheels starting at 10″/12″ (254 mm/ 305 mm) diameters. Knobs must be deep enough to be gripped well. Generally this means a depth of slightly less than 1″ (25 mm) : 1″ (25 mm) being the mean length of a first phalanges section of a finger. (Illustrations **64a**, **b**, **c**, **d**, **e** and **f**.)

Hand or power grip $1\frac{3}{4}$″ (44 mm) diameter. (Illustration **64f**.) Light finger control $\frac{3}{8}$″ (9 mm) to $\frac{5}{8}$″ (16 mm) diameter. $1/1\frac{1}{2}$ lbs (0·45 kgs/0·68 kgs) loading. (Illustration **64c**.) Single push button control $1\frac{1}{2}/2$ lbs (0·68 kgs/0·9 kgs) loading. (Illustration **64b**.) Multiple switches. Power hand grip $1\frac{3}{4}$″ (44 mm) diameter. (Illustration **64f**.) Toggle switches. 1″/2″ (25 mm/51 mm) high $\frac{1}{2}$ lb (0·23 kgs) loading. (Illustration **64a**.)

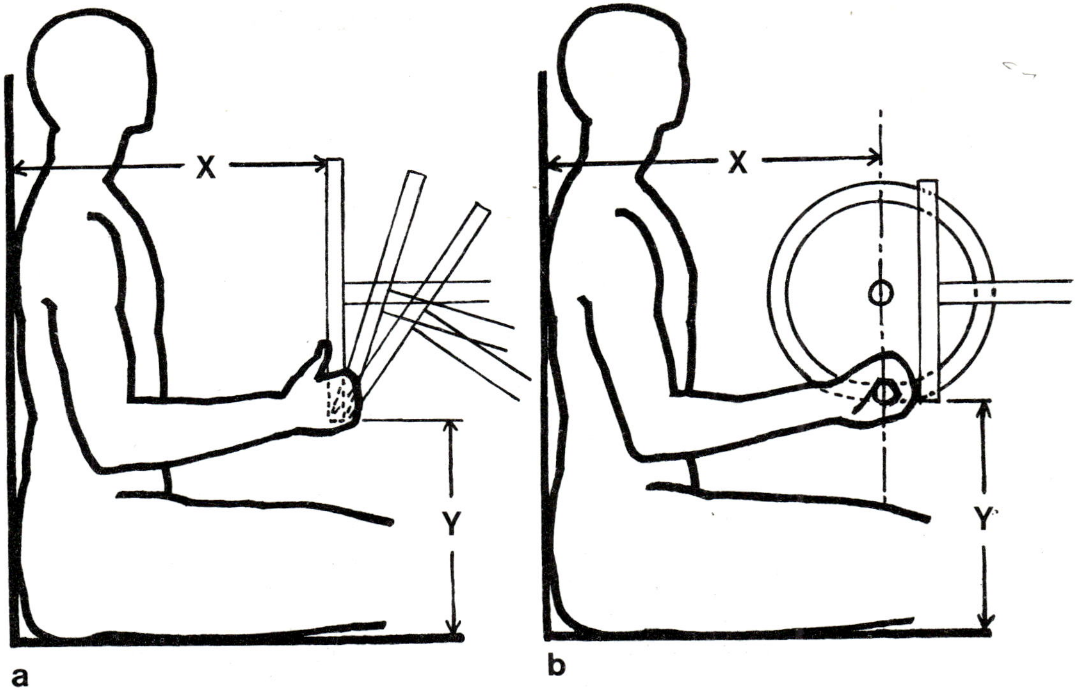

Percentile values for hand dimensions in inches and millimetres.

	5th in	5th mm	50th in	50th mm	95th in	95th mm
Hand length						
Male	7	178	7·5	190	8	203
Female	6·5	165	7	178	7·5	190
Hand breadth						
Male	3·7	93·5	4·1	104	54·5	114
Female	3	76	3·5	89	4	102

Force exerted by squeezing the hand or handgrip

Male	5th percentile	55 lbs (24·9 kgs)
Female	5th percentile	50 lbs (22·7 kgs)

65a and b

a Suggested clearances for handwheels.

X = 15″ (381 mm)
Y = 9″/10″ (229 mm/254 mm)

b Suggested possible positioning for a small crank wheel. x and y are still 15″ (381 mm) and 9″/10″ (229 mm/254 mm) respectively. Two positions for crank wheels are shown. special heavy duty clothing might call for revision of these clearances.

So that a safe maximum for a spring loaded grip or brake might be something in the region of 30 lbs (13·6 kgs).

Forces of 90 lbs (40·8 kgs) to 110 lbs (49·9 kgs) can be exerted by squeezing the hand, but these forces are approaching 95th percentile values.

Joysticks

Joystick operation involves movement of the arms and hand near the body and thighs. The loading for this type of control for 5th percentile operators is 15 lbs (6·8 kgs). The advantage of the joystick is its range of movement in an area in front of the body: their exertable force taking second place. Taking these facts into account 15 lbs (6·8 kgs) would be a good joystick loading. Lateral and fore and aft travel should be about 6″/7″ (152 mm/178 mm) in each direction. These distances relate well to 5th percentile arm reaches and the possible positioning of pedal controls. Maximum trunk depth, however, must be considered when positioning a joystick. Therefore adjustable seating would offer the best solution for fitting an operator to this type of control.

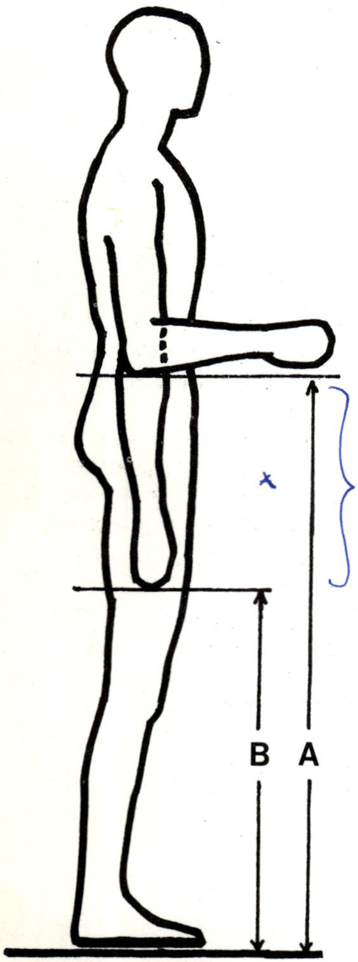

66
Switches and handles
General-purpose switches and door handles are convenient if placed somewhere between elbow height A and finger tip height B. Taking the two smallest values for these distances we can find we have for women A 34″ (559 mm) and B 24″ (610 mm). This gives a band of 10″ (254 mm), which is 24″ (610 mm) above floor level, in which switches and handles can be placed, and which is convenient for the tallest person.

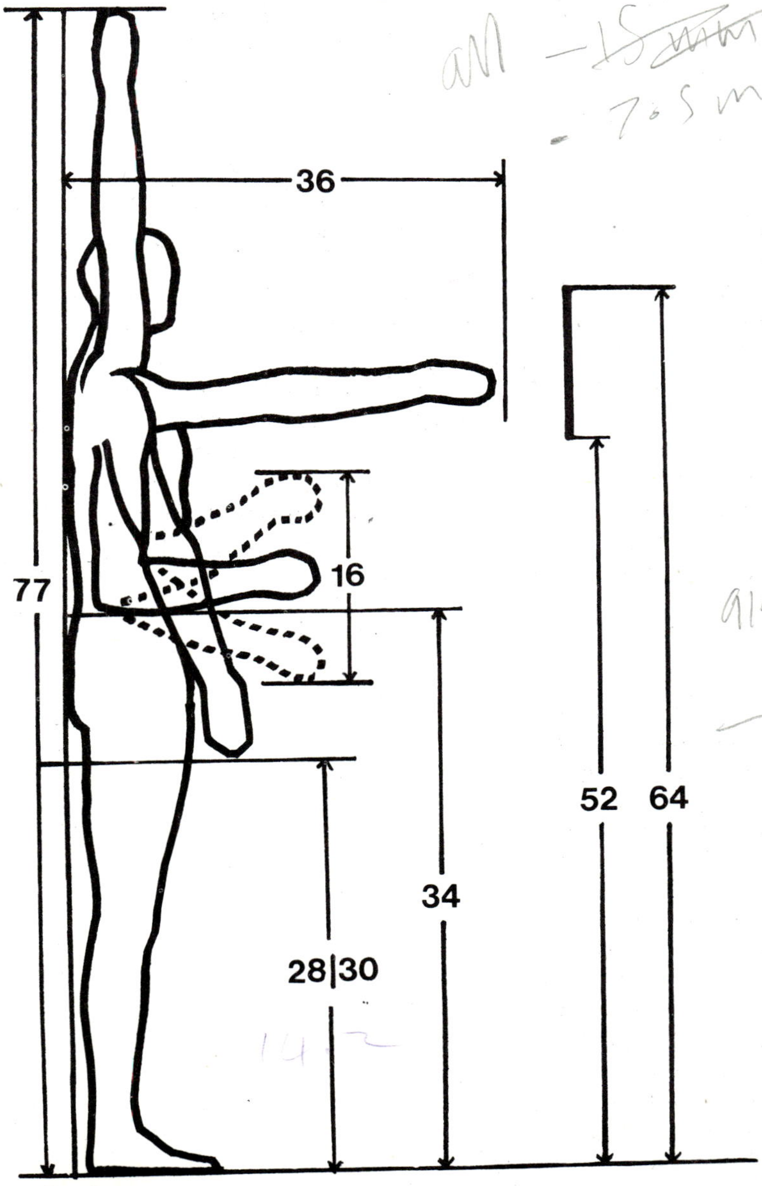

67
Reach and Stretch Dimensions

77″ (1956 mm) is the maximum stretch reach above the head for the 5th percentile, it would also allow an inch or slightly more clearance for standing height for a 95th percentile man. 36″ (914 mm) is the minimum distance that would allow all arm lengths to move unimpeded at full stretch. 52″/64″ (1321 mm/1626 mm) is the position for controls and small dials that could be seen clearly by most men and women. 34″ (559 mm) would allow clearance for most elbow heights. Hand height is difficult, but 28″/30″ (406 mm/762 mm) would be convenient for many people; something like 50 lbs/60 lbs (22·7 kgs/27·2 kgs) pressure can be applied by the hand from this position. The 16″ (406 mm) range of hand position in the perpendicular plane is the maximum in which the hand can be manipulated without moving the elbow forward.

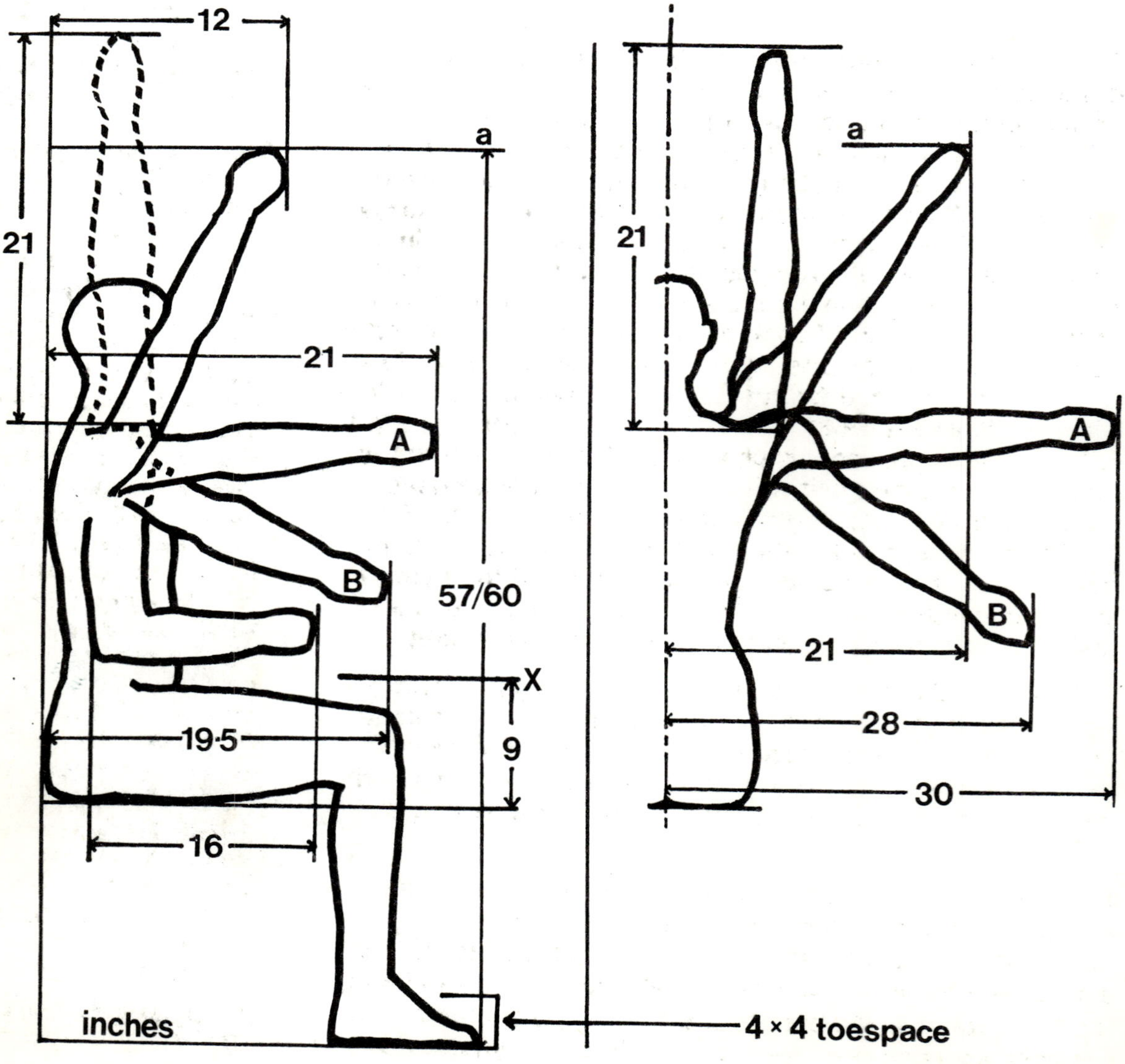
12
21
a
21
A
B
57/60
X
9
19·5
16
inches
a
21
A
B
21
28
30
4 × 4 toespace

Seats–Seating: General

Sitting is a means of changing posture and bringing rest. During the course of a day we continually change posture to reduce fatigue. Probably, of all the possible postures, the most effortless is the fully reclining semi-foetal position often adopted in sleep.

Sitting can be a tiring and painful business on a poorly designed seat. A good seat should allow for movement or a change in the sitting posture; there must be space for easement to maintain the best sitting posture for a lengthy period, but there should be enough control from the seating surfaces to effect the relief of body weight and give a sense of stability. A seat needs to give rigid support but not rigid confinement. It should in particular support the thorax and pelvis and help to maintain the angle of the spine between. Therefore, the design of back-rests is important. A low seat is better than a too high one, but no seating should be so low that legs are stretched far out in front of it without leg or foot support. (Illustration **69b**.) A working chair of correct height will allow leg movement backwards with the feet flat to the ground. A too-high chair will cause unnecessary pressure under the front of the thighs, and give pain and discomfort, for the soft tissues of the thigh is not meant to be a weight bearing area, and its compression can only lead to traumatic effects. Sprung upholstery or a cushioned surface can be good, but it is not a substitute for incorrect chair height.

Quite involved adjustable seating is used in industry and health-care work, and it is always necessary where seating dimensions are critical. Adjustable head, back, arm and foot rests can all lend support to a more effective use of the body when seated. Adjustable seats and supports can supply a controlled mobility.

Collections of measurements may be the beginning of the design process for seating, but only practical trials of prototypes will suggest the necessary modifications that make a successful design. It is useful to take past seating solutions that have reference to current problems and view them together, superimposed over one another by photographic

68

Minimum reach distances for hands clasping controls based on 5th percentile figures. A is reach at shoulder height. B is reach at waist height. X is the underside of a work surface and 9″ (229 mm) is given as the minimum clearance for the thigh. 16″ (406 mm) is given as the maximum clearance for the forearm. In general the 5th percentile reach distances, which are nearly the shortest reaches, should always be used. All critical controls should be placed opposite the operating limb. Forward reaches can be improved by up to 6″ (152 mm) by bending the trunk forward. This however decreases reach distances above the head. Seat height has not been stated in this diagram, but it would be normally estimated as varying from 15″ (381 mm) to 18″ (457 mm). The seat-height/work-height relationship is much more important than seat-height/floor-relationship. Adjustable seating and adjustable foot rests can be used to maintain optimum seat height/work height relationship.

a on both diagrams is 57″/60″ (1448 mm/1524 mm).

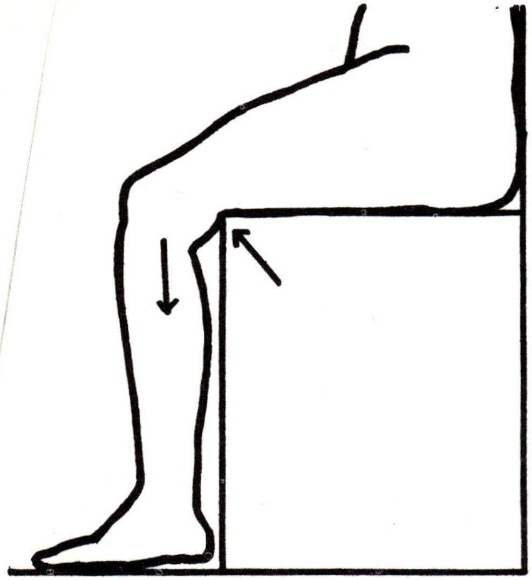

a

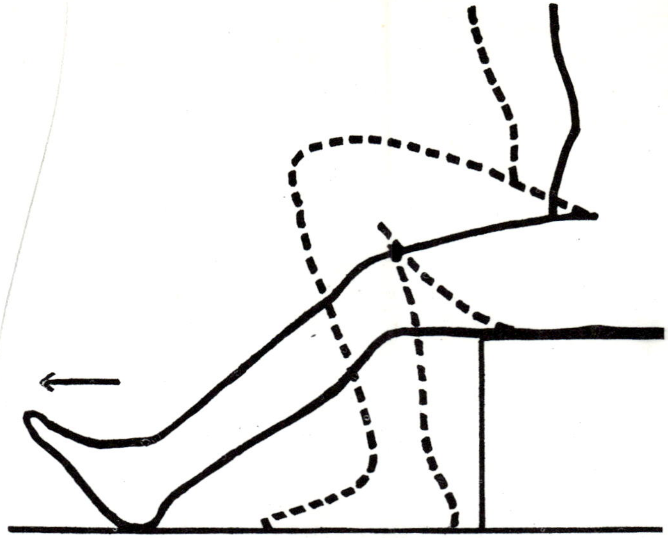

b

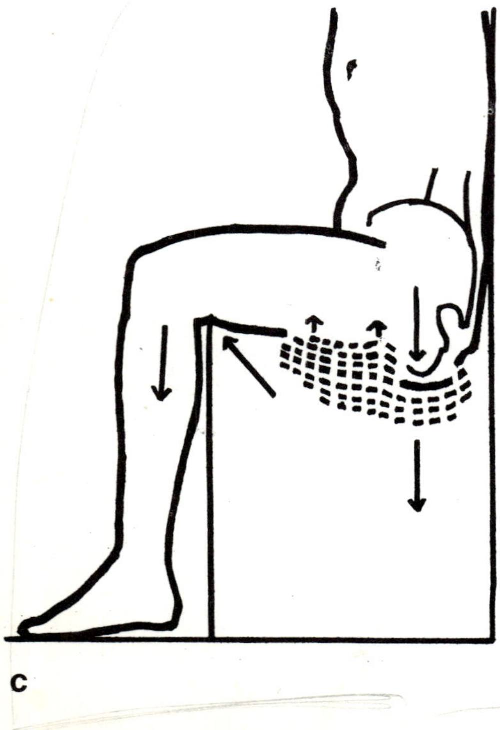

c

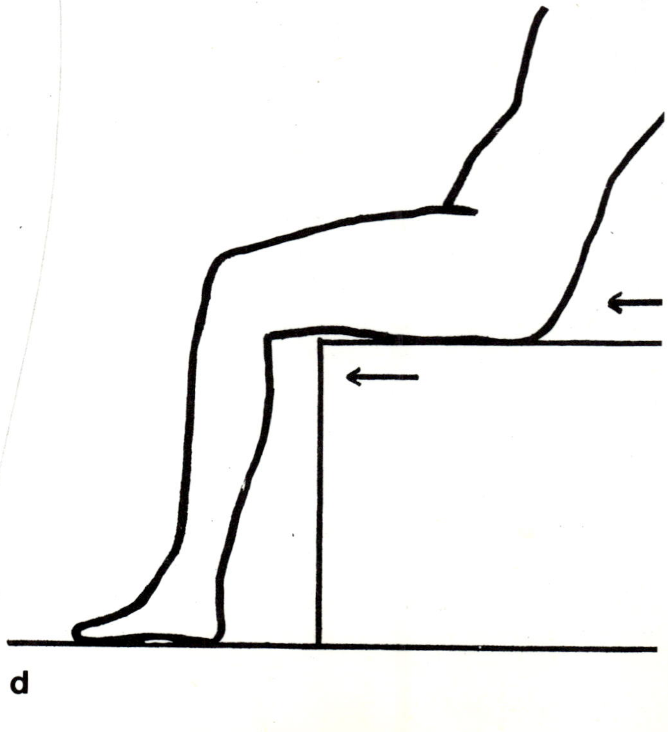

d

methods. Critical measurements in the current problem can be tried against them. Photography can also supply a rapid record of the varying postures a trial population naturally adopts when using the prototype. It will be found that the study of the variation of posture by trial users will suggest modifications to features and dimensions.

There is a wide range of types possible in seating design, and with the absence of a specific demand the possible use of a new design should be carefully considered.

Seating: Height
Of all the dimensions of seating, seat height is one of the most important. A chair height must not be so high that tissue in the distal and posterior region of the thigh is compressed and the seat front edge forced to act in the manner of a tourniquet to the blood supply of the legs. (Illustration **69a**.) Every seat height should be able to fit the shortest length of any lower leg. An acceptable height fit should allow a small space under the distal portion of the thigh as a hanging space for soft tissue, when a seated person would have their knees at right angles and feet flat on the ground. (Illustrations **70a** and **30x**.) By reason of the effect of tourniquet a tall person can find more comfort in a low seat than a short person can from a high seat. (Illustration **69a** and **b**.) In general if a seat is too high a person's body will slide forward and any special planning for the depth of a seat or its back-rest will be lost. If a seat is too low the seated posture will assume a forward crouch with most seating benefits lost. As anthropometrically there is such a wide range of lower limb dimensions, for critical situations in industry and commerce adjustable seating provides the best solution. Adjustable seat height should offer a range of from $14\frac{1}{2}''$ (368 mm) to $18\frac{1}{2}''$ (470 mm) for men, and from 14″ (356 mm) to $17\frac{1}{2}''$ (444 mm) for women. (Illustration **71**.)

Seats are often found to be 18″ (457 mm) high, but this is too high for the majority, and too high for a seat for a working position for long periods. 17″ (432 mm) has been recommended, but this is still a height which is excessive for women.

69

a Seat too high. Front edge of seat surface cutting into back of lower thigh.

b Seat too low. Legs slide forward unsupported without stability at the foot. The dotted line shows low seat crouch.

c Upholstery too soft. Weight not taken by ischial tuberosities. Front edge of seat surface cutting into back of lower thigh.

d Seat surface too long. Seat edge cuts into back of knee and sitter slides forward as a consequence, the back rest then being unused.

70

a A slightly tilted seat base and angled back rest to help comfortable 'settling' in a chair.

b Angled back rest assisting the lumbar curve to maintain its natural curve or 'arch'.

Insert (**b**) shows the straightening of the lumbar spine when there is an angle of 90° or less between the seat surface and the back rest.

A reasonable fixed seat height for men would be $16\frac{1}{2}''$ (419 mm), but for a woman $15\frac{1}{4}''$ (387 mm). 16″ (406 mm) offers a good compromise and is, in fact, a good height for adults, and the optimum sitting height. For general leisure purposes seat height can be reduced still further to something under 16″ (406 mm) with a minimum at 15″ (381 mm). Where the lower leg is angled forward and the seat is tilted, as in most driving positions, the height of the seat can be 14″ (356 mm) inches or less. But for quality driving a fully adjustable seat which offers each individual the best compromise between reachability of controls and maximum vision is best.

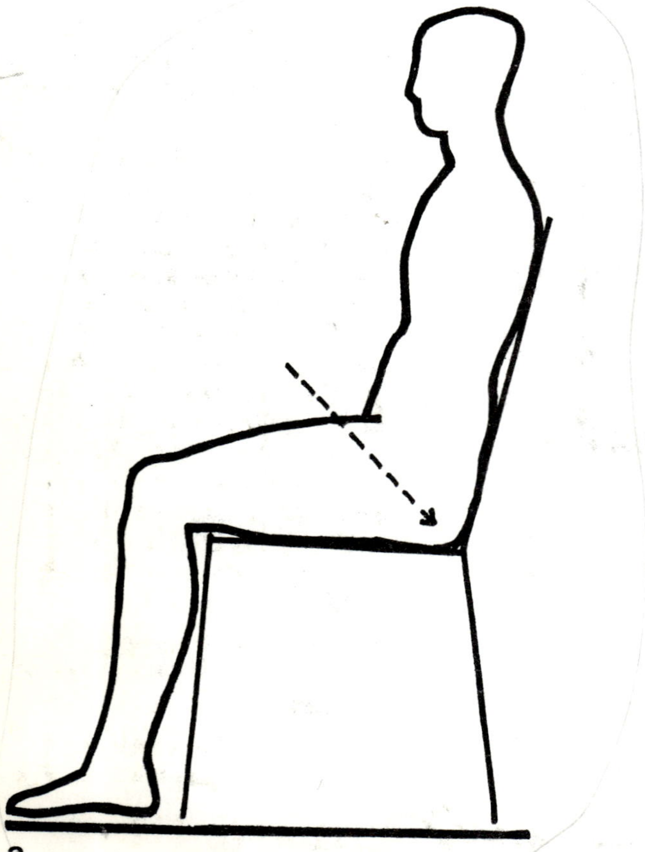

a

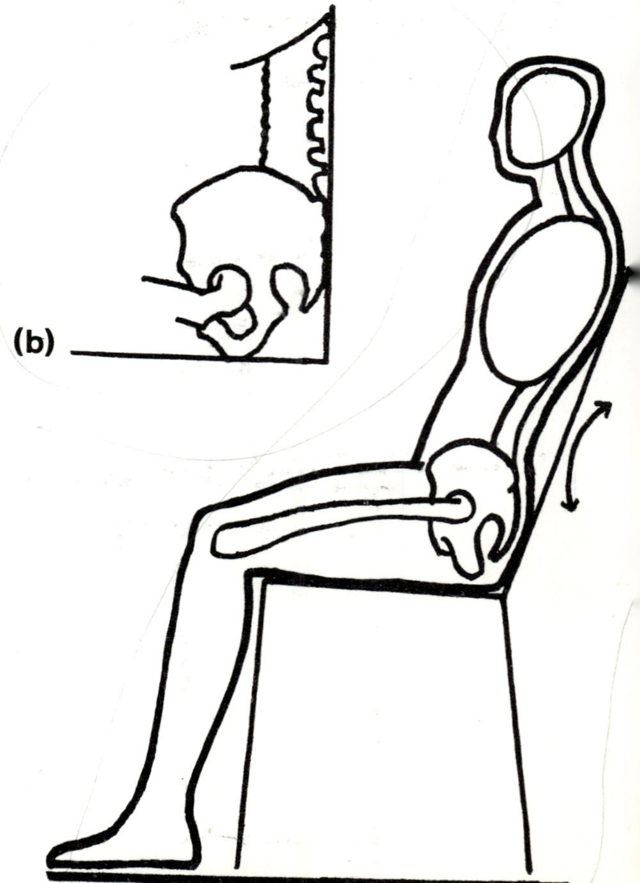

b

71
The adjustable chair

The great value of the adjustable chair is that it can be altered to allow for the very small differences required in a chair from person to person. For important and high-level quality operating sitting positions, it cannot be bettered. When it is considered that it has been found than an eighth of an inch addition to seat height could give discomfort it can be understood how an adjustable chair might be able to take care of individual seating troubles that anthropometric study might fail to do, and that our sitting habits require.

A Arm rest height. 8″ (203 mm) to $8\frac{1}{2}$″ (216 mm). If it could be adjustable a range of from $7\frac{1}{2}$″ (190 mm) to 10″ (254 mm) is recommended.

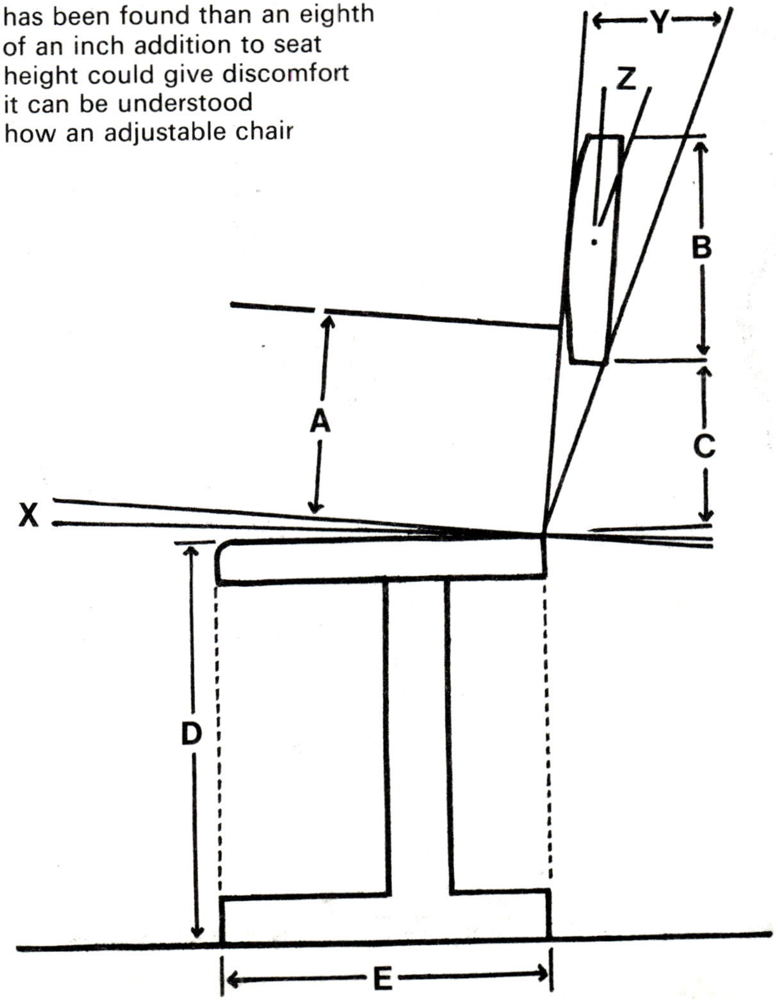

B back rest height. 4″ (102 mm) to 6″ (152 mm) for men. 8″ (203 mm) for women. It should be able to tilt independently z from angular back rest movement Y. Its surface to be slightly convex.

C Back rest height from seat surface 6″ (152 mm) to $7\frac{1}{2}$″ (190 mm) for men 5″ (127 mm) to $7\frac{1}{2}$″ (190 mm) for women. Height adjustable if possible.

D Seat surface height (14″ (356 mm) to 19″ (483 mm). A wide range which 'fits' most lower leg sizes.

E Seat length or depth. $13\frac{1}{2}$″ (343 mm) to 15″ (381 mm). This presents a dimension which should be comfortable for a wide range of men and women. It should not cut the back of the knee. It could be extended to 16″ (406 mm), but then the tolerance for a firm topped working chair has been made very fine. Reduced to 12″ (305 mm) it could give an operator greater mobility, but is then something of a momentary perch and would require foot rests.

X Angle of seat surface. From 3° to 5° or from 0° to 5° degrees.

Y Angle of back rest. From 95° to 115° from seat surface.

Seating: Foot-Rests

Foot rests must be used where the seating position carries the whole body above normal sitting height level. If they are used for this purpose or any other they should always allow the angle between the lower leg and the base of the foot to be normal or about 90° to 100°. (Illustration **72**.) Either when used as a built-in fixture on a piece of apparatus or a chair, or separately as a foot-stool, if the surface is greater than 15° from the horizontal a heel stop should be used. (Illustration **72b**.) The overall surface needs to be large enough to rest the whole foot.

72

a X angle of foot rest surface to lower leg. 90° to 100°.

b If the angle of a foot rest to the ground is 15° or more at Y, then a heel stop should be provided.

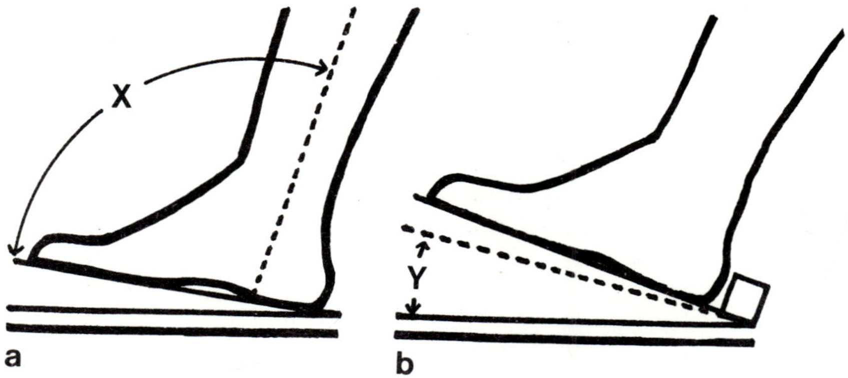

Seating: Arm Rests

Arm rests are often used as the bases for arm leverage in working positions, or for getting in and out of a chair. The critical distances are those between the arm rests, and between the seat and the arm rests. Arms should be able immediately to find the rests without sideway search, and the arm rest height should be able to accommodate upper arm height comfortably without levering the trunk out of the chair. (Illustration **73**.) Between extreme figure types there is no ideal height for arm rests. In critical situations an adjustable rest would have to be designed. The arm rest can be very useful where an operator has to use fine and sensitive finger and wrist operations unhampered by arm weight. Like seat surfaces they should not be too soft or too smooth.

A useful distance between arm rests is 19″ (483 mm). A useful workable height is eight inches, but this might have to be adjusted to eight and a half inches if padding rests were used. A minimum and maximum range for arm rest heights is from 7½″ (190 mm) to 10″ (254 mm).

Seating: Width
This dimension only matters when a minimum allowance has to be thought of for hip width and the spread of the buttocks, of for a shifting of position. Having looked at these factors a comfortable width is 18″ (457 mm), and nothing less than 17″ (432 mm) is tolerable for lengthy periods. In special circumstances space would have to be allowed for clothing, particularly where seats are confined by arm rests. (Illustration **74c**.)

Seating: Composition (Hardness or Softness) and Shape
Common experience and testing has shown people seated in softer seats give a better performance; that is, greater efficiency, for a longer time. Anatomically it is the ischial tuberosities that carry the body's weight in the seated position. (Illustration **70b**.) The tissue region over these bony protuberances has a superior blood supply to other tissue in the buttock formation. Therefore, because of the foregoing reasons, although seat surfaces need to be soft, they must at the same time be able to exert counter-pressure and not submit weakly to the body's weight. A depression of ½″ (13 mm) in a padded seat is enough. More than that does not allow the average ischial tuberosity to take the majority of the weight and the surrounding tissues become too compressed. (Illustration **69c**.)

For short-period sitting a seat surface can be completely firm, providing space is left for shifting the body position. Allowance for adjustment of position delays fatigue in sitting.

Sculptured seat surfaces with slight depressions carved to take the bony protuberances have been tried. There are so many buttock formations anthropometrically that seat designs based on this idea can only expect limited use.

73
If arm rests are too high their purpose is defeated for they tend to lever the sitter out of of the chair, or bring unnatural pressure on the shoulder joints.

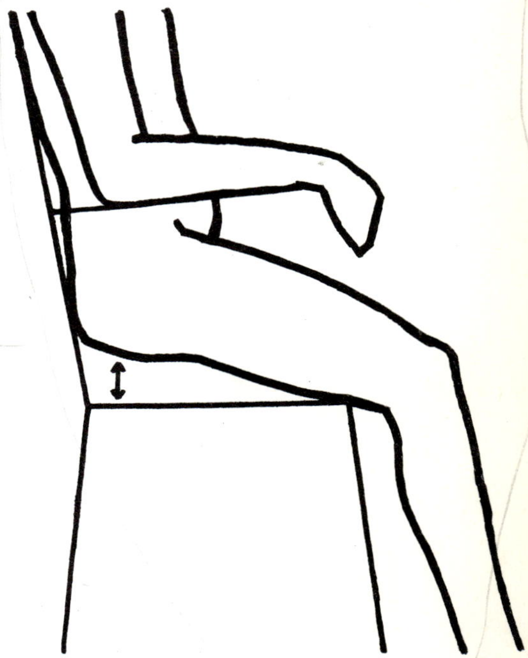

74
The consideration of space in seat design.

a Seat back width at shoulder level or above for leisure and lounge seating 20″ (508 mm) or more. 20″ (508 mm) incorporates shoulder widths of the 95th percentile for men. 21″ (533 mm) is the 99th percentile value. Curvature slight for leisure. Radius of curve two and half seat lengths. Seat back width at 14″ (356 mm) high 15″ (381 mm) to 16″ (466 mm).

b Seat back height. 14″ (356 mm) high for utility, 20″ (508 mm) to 25″ (635 mm) for shoulder support. (95th percentile value for men's shoulder height is 25″ (635 mm). 22″ (559 mm) would be maximum for driving seat where all round vision is required. 35″ (889 mm) for back and head support.

(b) Allow for movement of legs under chair for alert and working seating. Also allow for 4″ (102 mm) between front of chair and heel (X).

c Seat width. 18″ (457 mm) for comfort and where there are enclosed arm rests it is a minimum. 19″ (483 mm) better. Arm rest widths 4″ (102 mm) or more. (Y)

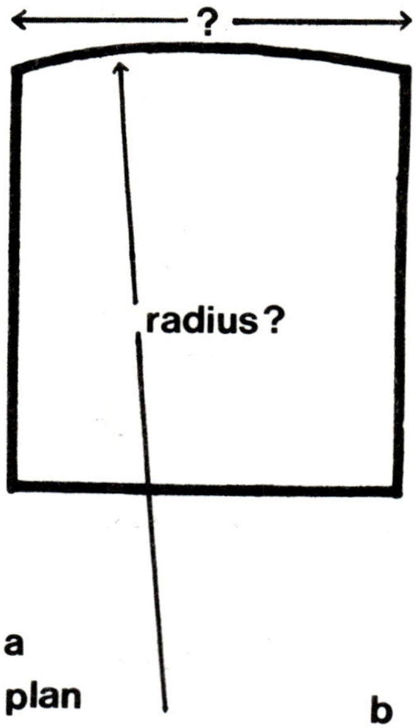

a
plan

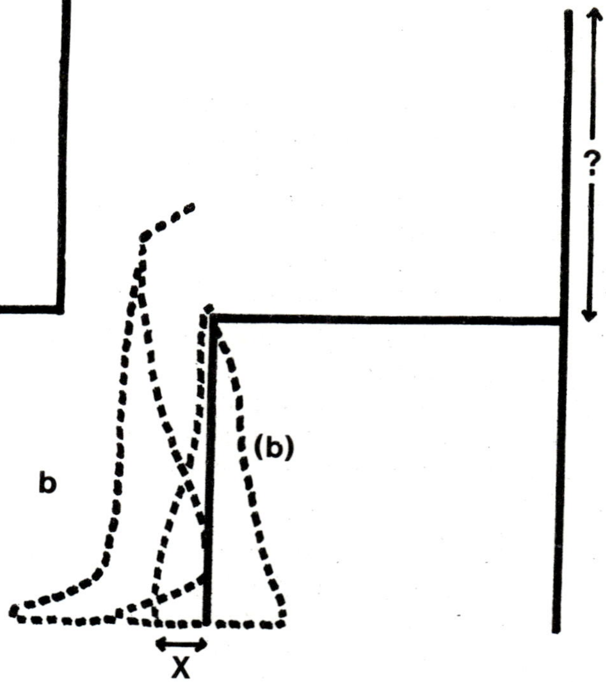

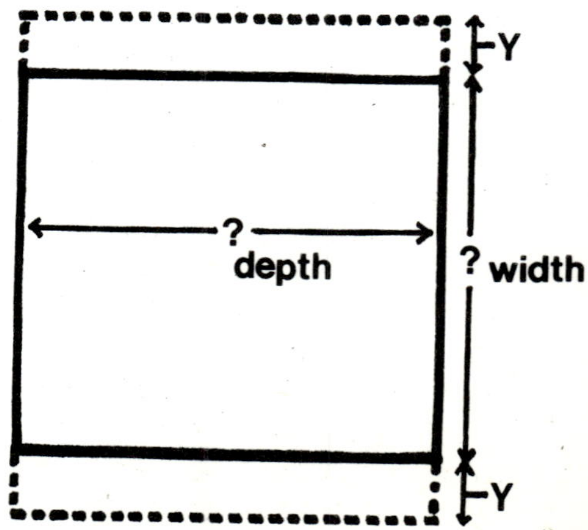

Seat depth. 15″ (381 mm) is safe. 20″ (508 mm) for lounge or leisure, but only when front of seat height is reduced to about 14″ (356 mm).

Seat surfaces should not be too highly polished. A textured or rough surface is best.

A rounded seat front edge is also good.

Seating: Depth
The depth for a seat does not present any great problem. The minimum back of knee to buttock measurement on a woman is about 16″ (406 mm); the maximum of the same measurement on a man is about 20″ (508 mm) or more. Taking the smallest measurement as a guide 15″ (381 mm) for seat depth would be suitable, and this is a measurement which in fact is often used. (Illustration **69d**.)

Chairs made from hard materials, or with firm padded surfaces, are better with depths of 15″ (381 mm) or less. In every case they would be improved if the front edge of the seat were rounded. With easy chairs depth can be generous as long as the upholstery is soft and the seat height is low.

Seating: Back-rest
Back-rests may be constructed at a varying series of angles from a seat surface, depending on whether an upright seated position or a reclining or semi-reclining position is required. (Illustration **75**.) A slightly tilted back-rest helps one to settle comfortably in a chair and prevents a gradual slide forward of the body. (Illustration **70a**.)

Without the tilt of a back-rest the lumbar curve is unnaturally flattened and strain is placed on the intervertebral lumbar discs and ligaments. (Illustration **70b**.)

An angled back-rest also assists the force of gravity to settle the users body into the chair and keep in such a position that the lumbar support section of the back-rest is used to the full. There are divided opinions on the use of a tilted seat base to further aid the cause. Probably in an upright working position adopted in offices a sloped seat would only be a disadvantage. A tilted seat base has an advantage in a semi-reclining position where the angle of the back-rest tends to push the trunk forwards. The possible angles for a back-rest from the perpendicular in the sagittal plane, for the working upright position, varies from 5° to 20°. (Illustration **75a**.) For comfortable seating this angle could be increased to 35°, but this would then be combined with some angling of the seat base from some 5 to 7°. (Illustration **75b**.)

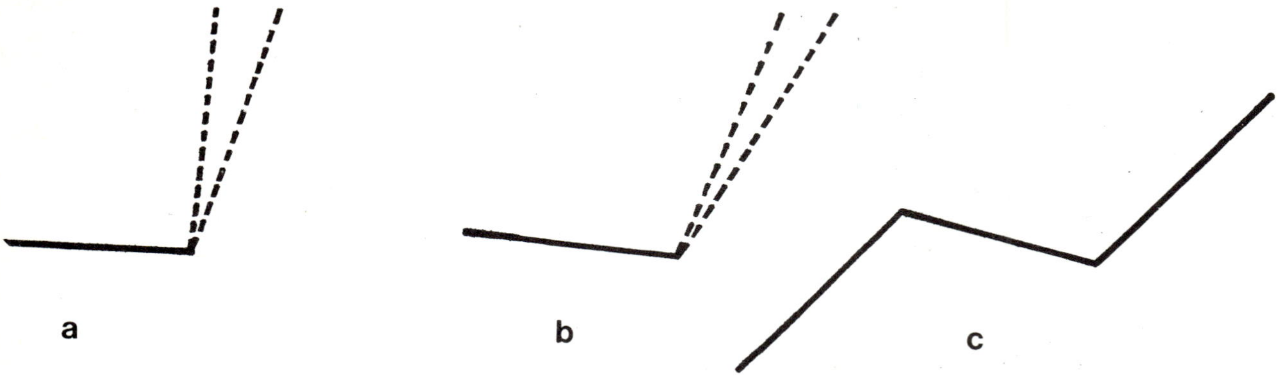

75
Back rest angles.
a Alert or attention sitting

b Every day, conference seating, relaxed travelling position. Back rest to horizontal level 110° to 120°. Seat surface angle 5°/7°.

c Reclining in comfort. Back rest to horizontal 135°.

Angles between seat and back rest, and seat and leg rest 120°.

The upright seated position is one that can be maintained for only a short time without support, and a back-rest can aid stability of the trunk and prolong the inevitable arrival of 'sitting-fatigue'. A good back-rest should allow the back freedom to be 'arched' occasionally and improve the lumbar curve. (Illustration **70b**.) Its shape should be designed in such a fashion that it supports the lumbar region of the back, and if it should be of sufficient height, allowance made for shoulder blade support. The support for the lumbar region is of prime importance. It should be raised clear of the sacral region, or else its effect will be lost, 6″/7″ (152 mm/178 mm) being a good height for its lower border. (Illustration **76**.) In an easy chair specially constructed for a woman this could be raised to 7½″ (190 mm). The height of the lumbar rest itself might be from 4″ to 6″ (102 mm to 152 mm); raised up to an 8″ (203 mm) height if specifically for a woman.

Space for the free movement of the legs under the seat also assists the maintenance of the lumbar curve during sitting, because backward movement of the legs relaxes the posterior muscles of the thigh and allows the pelvis and sacral spine

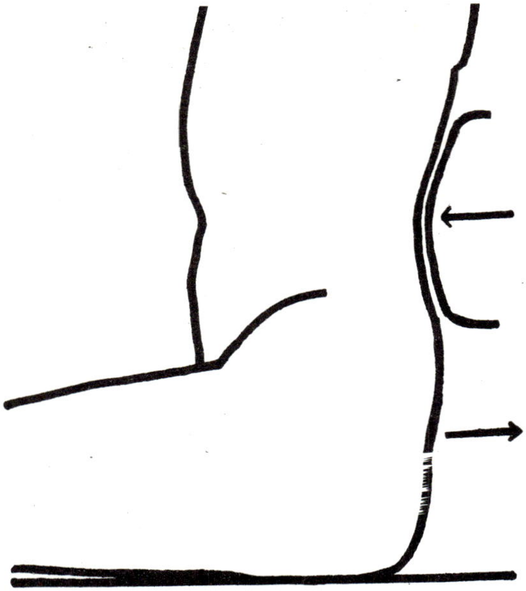

76
Back rest for a working position. Upper limit should be below shoulder blades so that movement of the shoulders and the arms is permitted. Thus the back rest becomes a lumbar spine support. About 12″ (305 mm) to 14″ (356 mm) wide gives free space for elbows.

The gluteal region should have space to move and press backwards under the backrest.

to rotate and maintain a normal relationship with the lumbar spine.

A fully comfortable and ideal back-rest might be one where the surface is moulded to accommodate the spinal profile to some extent. Such designs have been advanced from time to time with some success. The main advantage proffered by the idea of a continuously moulded back-rest, is that of being able to change the sitting posture from time to time whilst retaining back support. The shoulder support area in this type of design should make a minimum angle from the perpendicular of 15°. For complete relaxation the breadth of such a back-rest should be 20″ (508 mm).

Reclining
The reclining or semi-supine position is a favourite for relaxing the whole body. Keegan's angle of 130° between the seat and back-rest is the position which rests the muscles most. As the back-rest seat angle is increased, so there must be a corresponding increase in thigh lower leg angle. This is not a recommended position if vision and full head movement

is required. (Illustration **77**.) It prohibits watching television or normal level view. Vision can be improved by increasing the angle at the knees, but moving the head can be tiring as its rotation is decreased by the amount it has to be pulled forward and into an upright position. However restful, a reclining position is anti-social, in that the ceiling or the sky is the object of interest.

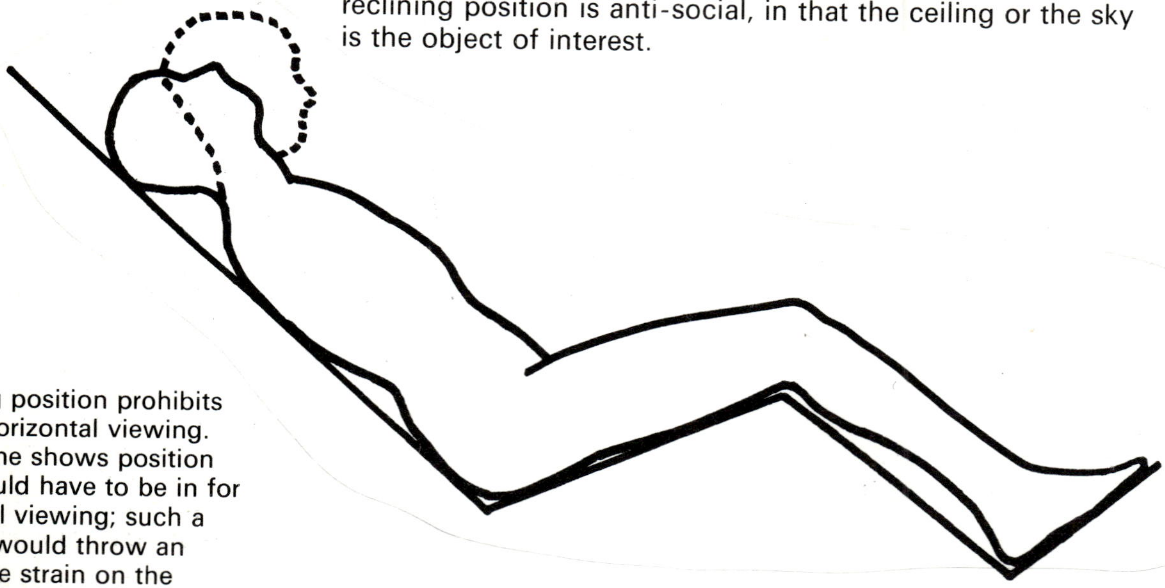

77
Reclining position prohibits normal horizontal viewing. Dotted line shows position head would have to be in for horizontal viewing; such a position would throw an intolerable strain on the muscles of the neck.

Viewing

If the head is not restricted the field of vision is large because slight movements of the head can appreciably widen the field of view. The conformable nature of the lens of the human eye always keeps the image in focus. The range of this agreement of the lens with the image allows us a comfortable minimum viewing distance of 16″ (406 mm) (the optimum is about 21″ (533 mm) up to a distance of some 20′ (6096 mm) or optical infinity.

The accompanying illustration demonstrates the cones of human vision and viewing angles. (Illustrations **78** and **79**.)

We gain a great deal of information by the use of our eyes, which can detect the very slightest tonal difference. We are able to identify a wide range of textures and solid forms,

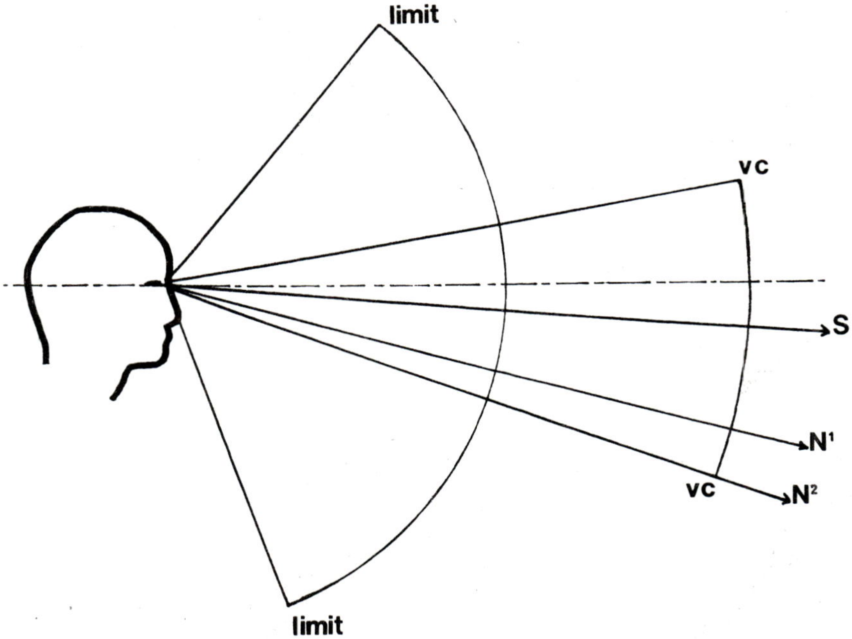

78
Diagram to illustrate lines of sight and elevation of visual field.

S is the standard line of sight and is 5° below the horizontal.

N1 is the normal line of sight when standing with the gaze undirected.
15° below horizontal.

N2 Shows a further drop of the line of sight when sitting with the gaze undirected. 20° below the horizontal.

We are a glancing-down animal by the construction of our eyes, which point below the horizontal. So that any horizontal scanning or looking up requires flexion of the neck and the use of muscles. vc is the visual cone 15° each side of the standard line of sight.

Limit. Marks an area 50° above the horizontal and 70° below which can be viewed by eye movement alone.

Control panel faces should be at 90° to the directed line of sight. So three possible control panel angles are: at S 5° to perpendicular, at N1 15° to perpendicular, and at N2 20° to perpendicular.

though an advanced visual skill must be developed by training. Apart from the head movements (Illustrations **79** and **81**.) given, and the various limits of vision, the eyes themselves are capable of movement. The rotation of the eyes adds to our ability to scan objects and the subtleties of their reflective surfaces.

5th and 95th percentile values for the heights of line of sight for standing and sitting positions in inches and millimeters

	MALE 5th	MALE 95th	FEMALE 5th	FEMALE 95th
STANDING	59·3 ins 1506·5 mm	67·5 ins 1714mm	55·2 ins. 1402mm	63·4 ins. 1670 mm
SITTING From seat Surface	29·3 ins 744·5 mm	33·7 ins 855·5 mm	26·1 ins 662·5 mm	30·7 ins 779·5 mm

79
Diagram to show plan of visual field and degrees of neck rotation.

45° is the degree of natural head movement.

60° is the degree of head movement achieved by conscious force.

vc visual cone 15° each side of medial plane, is the area of vision which can be scanned by normal eye movement.

Limit marks an area 30° each side of the medial plane, which can still be scanned by eye movement alone.

The maximum viewing angle can therefore be something like a sweep of 180° when eye and head movement are used together. That is 60° + 30° either side of the medial plane.

+ 90 indicates a "out of the corner of the eye vision", achieved by eye movement alone.

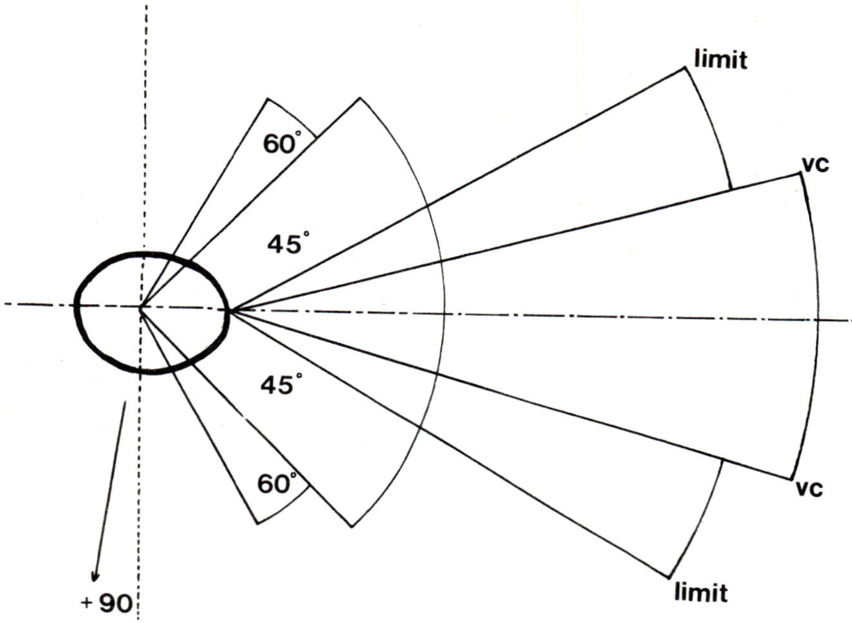

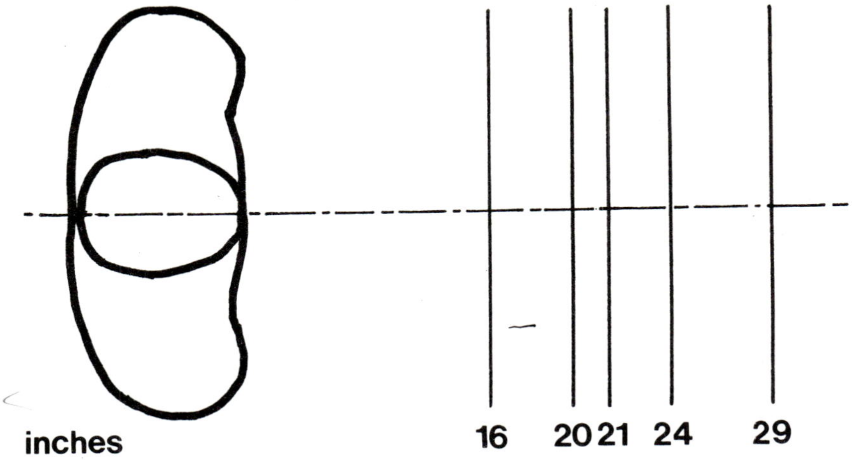

80
Diagram to indicate distances for the positioning of various types of information from the eyes.

16″ (406 mm) Minimum reading distance for print and instructions.
20″ (508 mm) Good reading distance for general instructions.
21″/24″ (533 mm/610 mm) Distance for dial readings. Dials should be kept very simple, with the minimum of markings. Good face diameters would be 2½″/3″ (64 mm/76 mm).
29″ (737 mm) Is the furthest distance for the display of information.

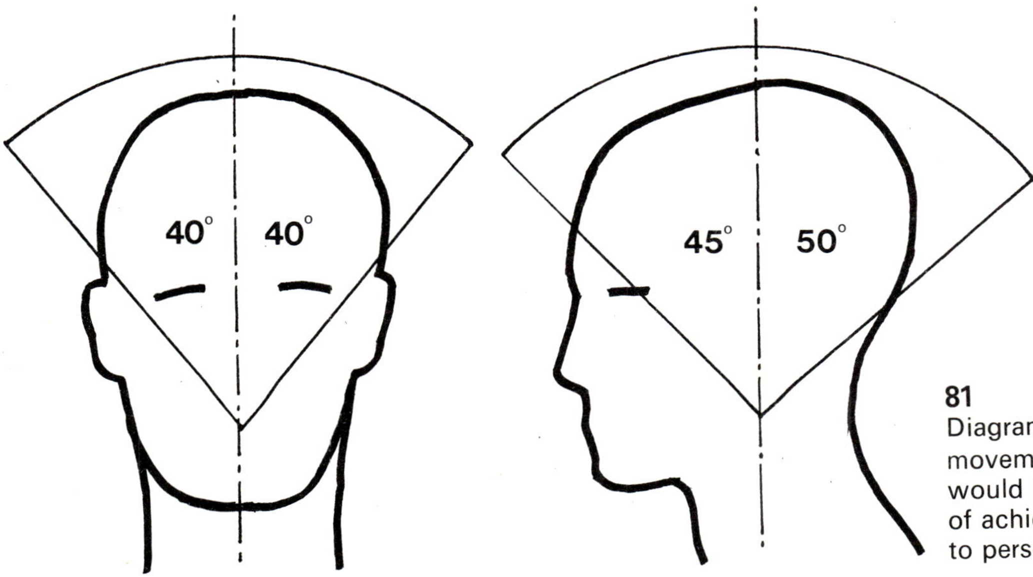

81
Diagrams to show flexion movements of the neck. There would be some variation of achievement from person to person within these limits.

Working Plane Heights

Heights for working planes for industrial use are impossible to define and place in a general table. Machinery heights vary greatly but it is advantageous to mount it lower rather than too high. Elbow height provides a measurement of the upper limit for working comfort. Heights lower than elbow height allow more force to be used. Individual types of manufacturing tasks will need greater or lesser amounts of space and height because of the varying range of body movements they require. Standing positions offer greater scope than seated ones. A greater range of heights can be used, along with larger controls and greater force, the only limitation then being the distance of instructions or gauges from the eye.

Height Limits for various functional planes

74″/78″ (1880/1981 mm) Door and window heights.

64″ (1626 mm)
60″ (1524 mm) Comfortable maximum height to reach to.

54″ (1372 mm)
52″ (1321 mm) Height over which most can see. Height for central displays or internal view over partitioning.

44″ (1118 mm)
41″ (1041 mm) Highest counters.
38″ (965 mm) Work benches: a good working height. Above the mean value for elbow heights for standing women.
36″ (914 mm) Popular work bench and sink heights. The counter height for women, and the height for standing ironing boards.

34″ (864 mm)
33″/32½″ (838/851 mm) Food preparation surface heights.
30″ (762 mm) Craft table tops and work room surfaces.
30″/28″ (762/711 mm) Average hand height standing.
29″ (737 mm) Maximum for desk tops. Can sit and lean. 29″ or 29½″ for dining room tables.

Maximum height for tops of controls or bottoms of wheels for seated operator.
28″ (711 mm) General table height.
27″ (686 mm) Minimum working sitting height for general tasks, desk or control desks. Sitting at ironing boards.
26″ Low working height. Typing tables.
25″
25½″ Accommodates all knee heights including value for 95th percentile for men.

24″ Minimum height for tops of controls or bottoms of wheels for seated operator.

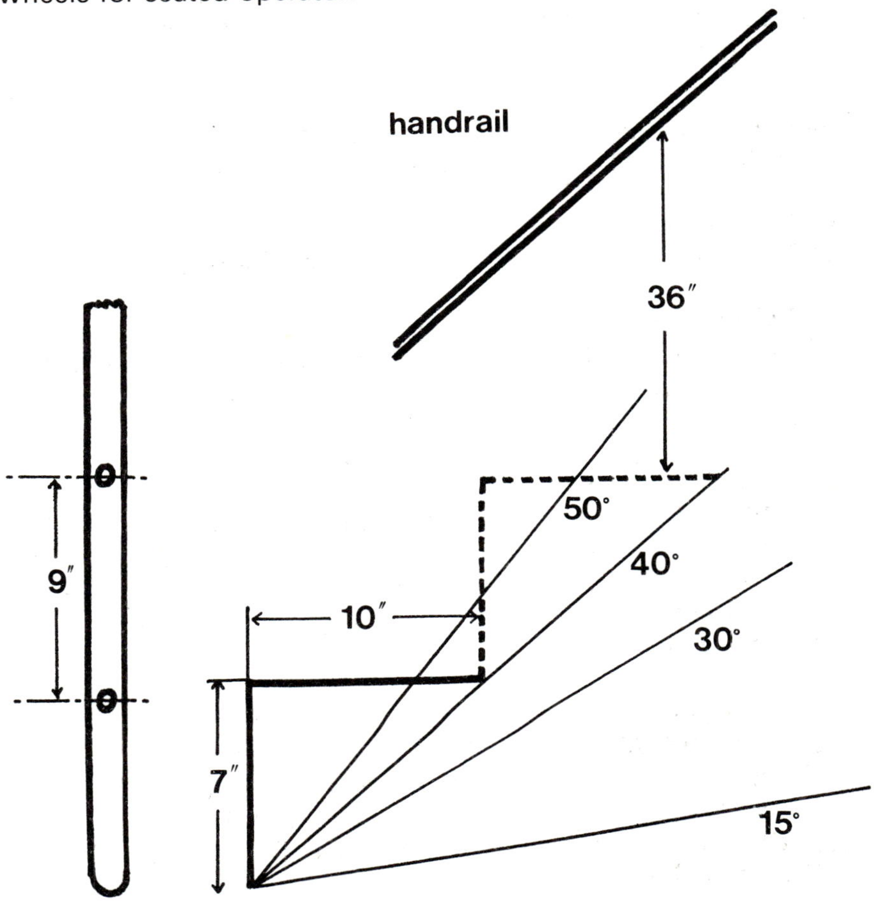

82
Stairs and Ladders

Stairs are comfortable set at about 40°. Beyond 50° they are really becoming ladders. Under 30° they might be replaced by a ramp or moving slope.

Stair rises can be about 7″ (178 mm), the variations being between 6″ (152 mm) and 8″ (203 mm); 9″ (229 mm) is too great. Stair treads can vary from 9½″ (241 mm) to 12″ (305 mm); they may be greater than this.

Stair widths can start at 2′ 9″ (838 mm). A 4′ (1219 mm) width or more is required when two people might be passing frequently.

Handrails should be between 34″ (559 mm) and 37″ (635 mm) high.

Ladders rungs should be spaced 9″ (229 mm) or so apart. 8″ (203 mm) or less is too small and anything more than 12″ (305 mm) can be tiring.

Openings to lofts need to be about 2′ (610 mm) square, the dimension which accommodates the shoulder width may be reduced to 20″ (508 mm) if necessary.

83
Clearances for Seated Operators
Minimum sitting clearances plan.

A 12″ (305 mm)

B 16″ (406 mm)

C 46″ (1168 mm)

D 50″ (1270 mm)
Person passing.

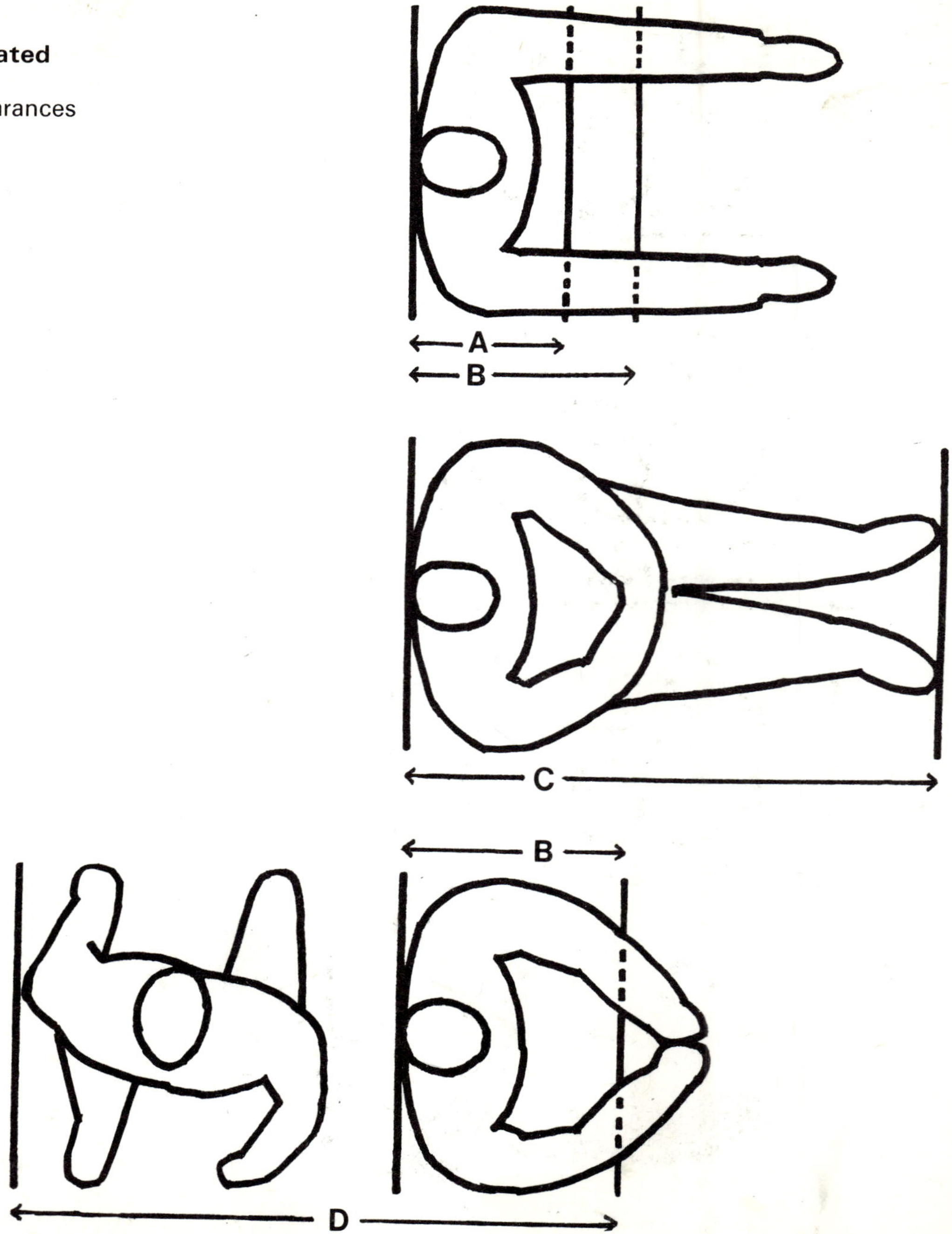

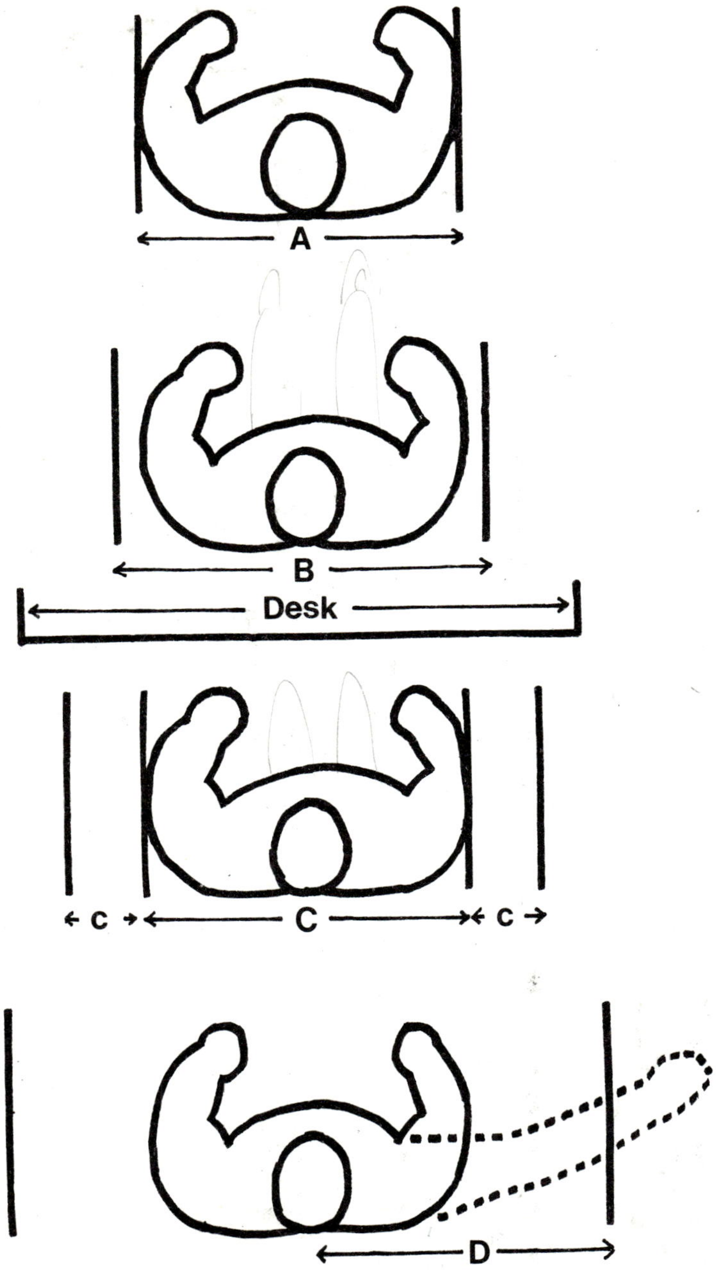

84
Minimum sitting clearances plan.

A 24″ (610 mm)

B 26″ (660 mm)
Eating 28″ (711 mm)
Theatre 30″ (762 mm)

C 24″ (610 mm)
c 6″ (152 mm)
Desk 42″ (1067 mm)

D Side counter 20″ (508 mm).

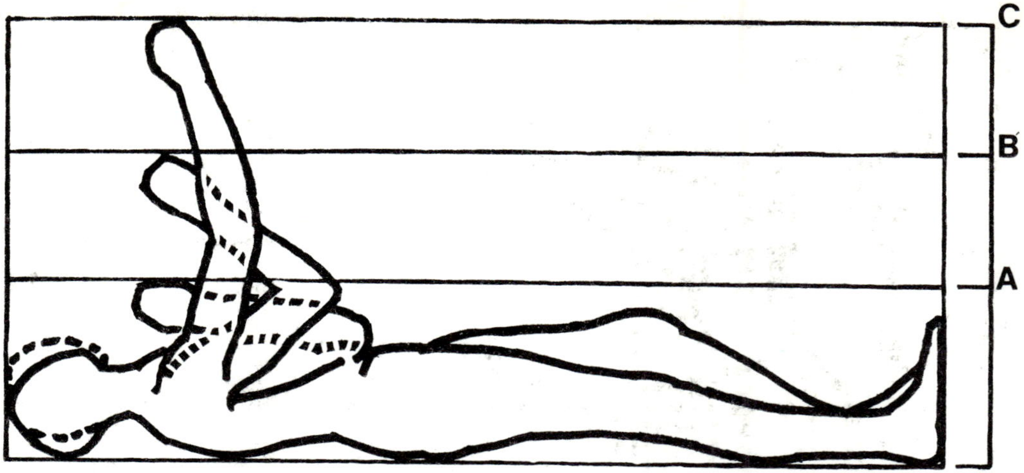

85
Inspection Clearances
Lying positions for inspection and repair. Minimum heights are given; the length is 76″ (1930 mm).

A Lying for inspection. Height 18″ (457 mm).

B Restricted space for small tools and minor adjustments. Power from elbow extensions not possible. Height 24″ (610 mm).

C Space for reasonable arm extension. 6″/8″ (132 mm/ 203 mm) length power tools could be used. Height 32″ (813 mm).

86
Minimum space allowances for the human body in a variety of postures.

A Standing. Space for reasonable arm extension. Depth 30″ (762 mm).

B Standing. Legs braced apart. Small power tools could be used. Depth 40″ (1016 mm).

C Kneeling. Depth 46″ (1168 mm).
c Height 48″ (1219 mm).

D Crawling on all fours Depth 48″ (1219 mm).
d Height 35″ (889 mm).

E Flattened crawling. Depth 58″ (1473 mm).
e Height 32″ (813 mm).

F Prone arm reaching forwards. Depth 96″ (2434 mm).
f Height 18″ (457 mm).

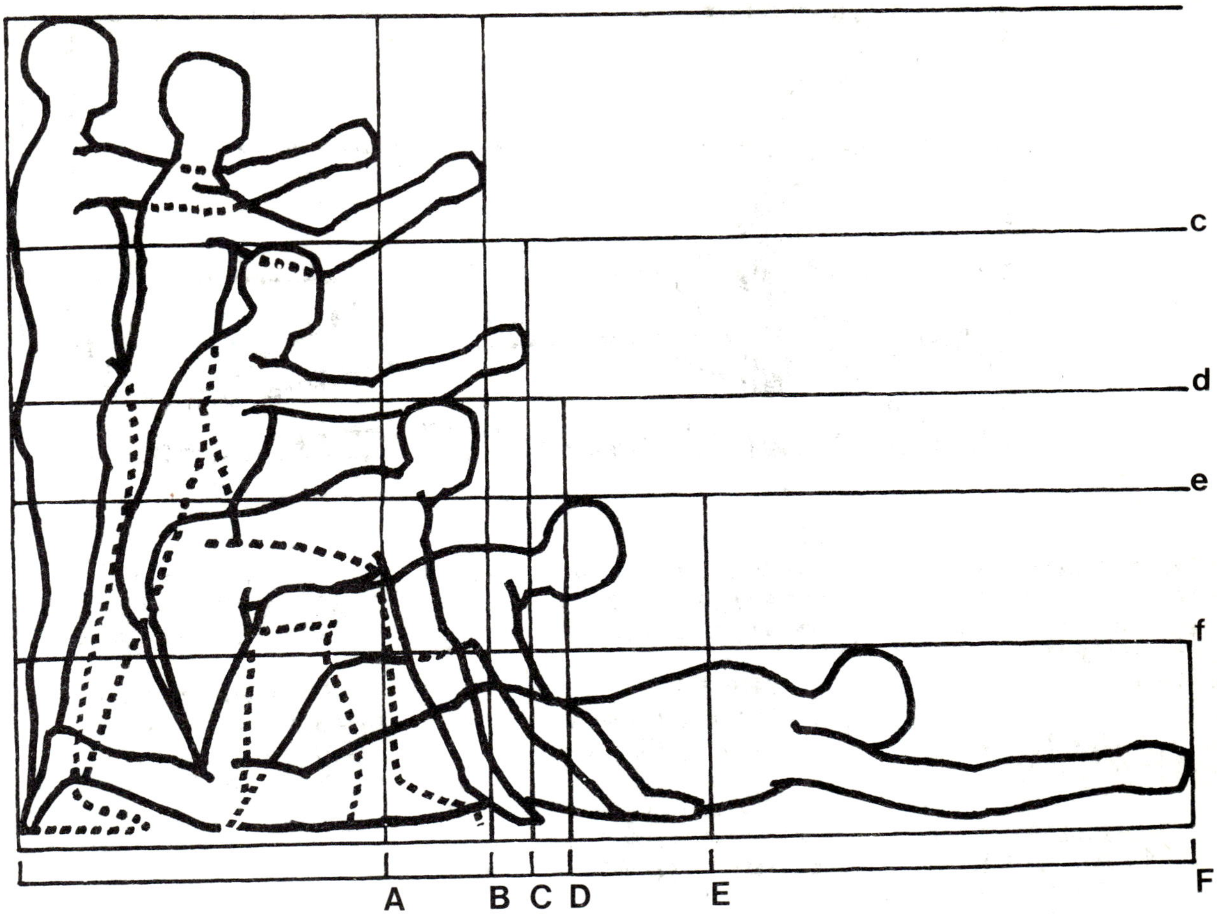
c
d
e
f
A
B
C
D
E
F

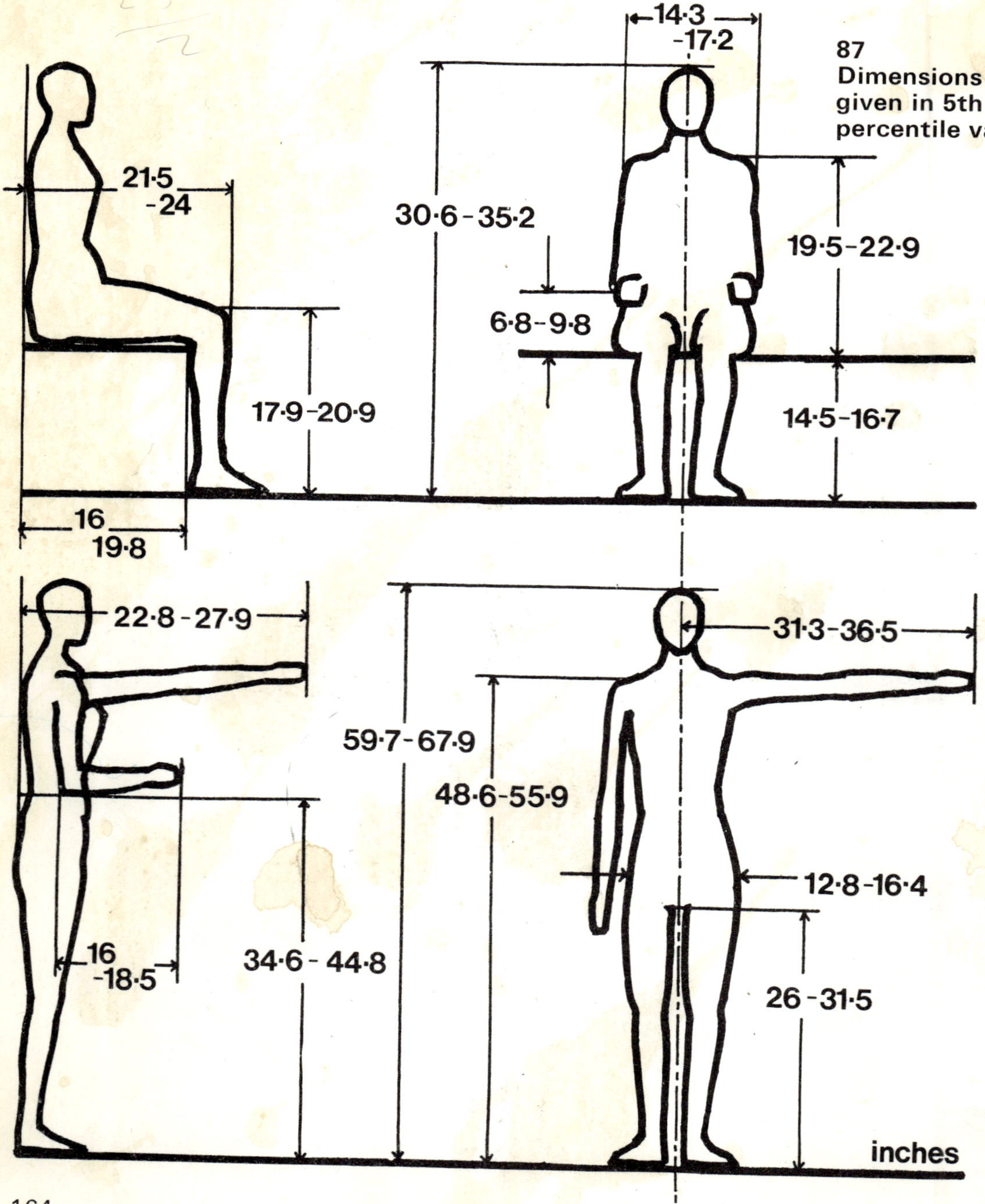

87
Dimensions of female body given in 5th and 95th percentile values.

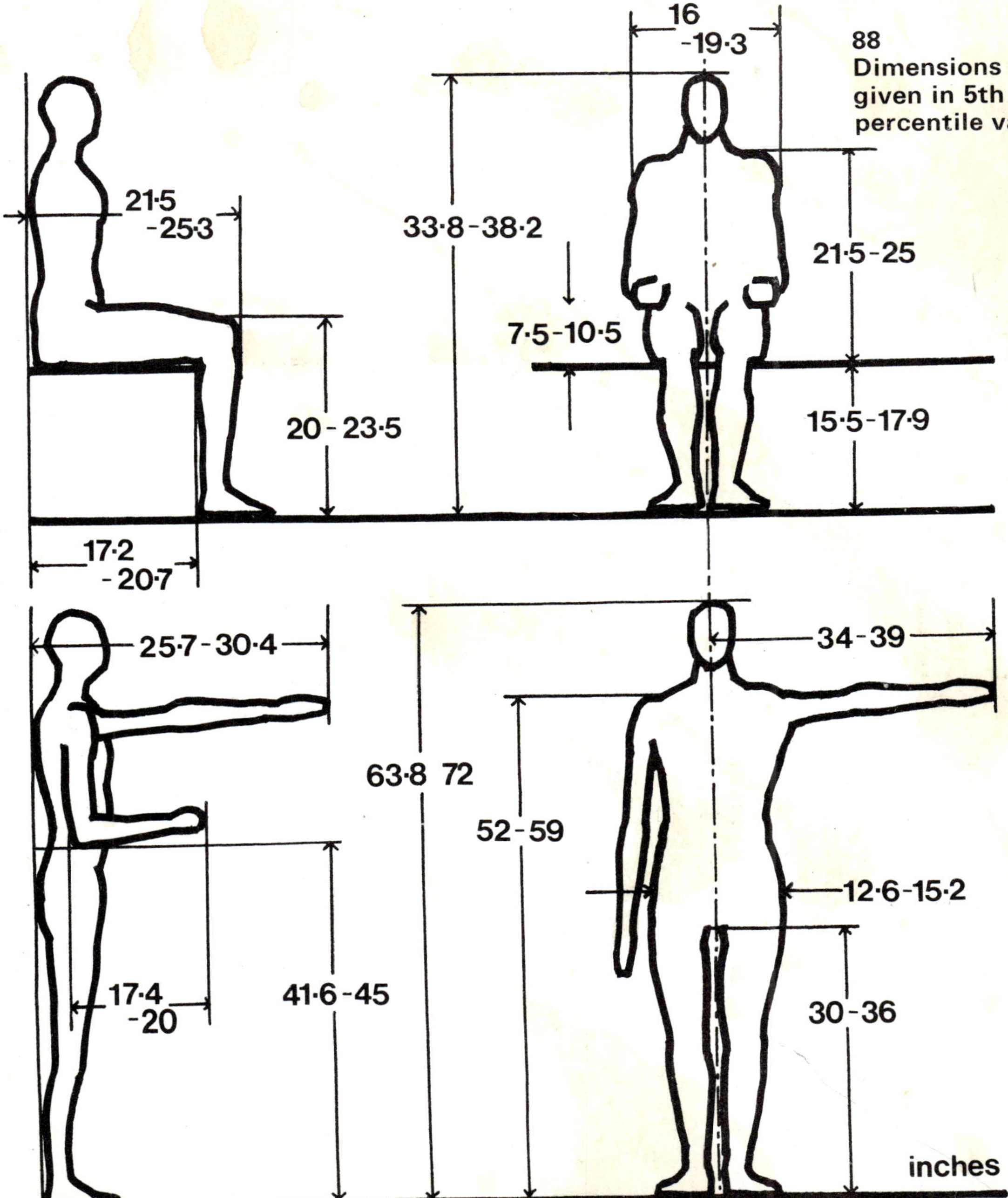

88
Dimensions of male body given in 5th and 95th percentile values.

Conversion charts for length, mass and plane angle

DECIMALS OF AN INCH TO MILLIMETRES *All values in this table are rounded to the second decimal place.*

inches	mm	inches	mm	inches	mm	inches	mm	inches	mm	inches	mm
0·00	—	0·10	2·54	0·20	5·08	0·30	7·62	0·40	10·16	0·50	12·70
0·01	0·25	0·11	2·79	0·21	5·33	0·31	7·87	0·41	10·41	0·51	12·95
0·02	0·51	0·12	3·05	0·22	5·59	0·32	8·13	0·42	10·67	0·52	13·21
0·03	0·76	0·13	3·30	0·23	5·84	0·33	8·38	0·43	10·92	0·53	13·46
0·04	1·02	0·14	3·56	0·24	6·10	0·34	8·64	0·44	11·18	0·54	13·72
0·05	1·27	0·15	3·81	0·25	6·35	0·35	8·89	0·45	11·43	0·55	13·97
0·06	1·52	0·16	4·06	0·26	6·60	0·36	9·14	0·46	11·68	0·56	14·22
0·07	1·78	0·17	4·32	0·27	6·86	0·37	9·40	0·47	11·94	0·57	14·48
0·08	2·03	0·18	4·57	0·28	7·11	0·38	9·65	0·48	12·19	0·58	14·73
0·09	2·29	0·19	4·83	0·29	7·37	0·39	9·90	0·49	12·45	0·59	14·99

INCHES TO MILLIMETRES (*1 inch = 25·400 mm*)

in	1	2	3	4	5	6	7	8	9	10	11
mm	25·4	50·8	76·2	101·6	127·0	152·4	177·8	203·2	228·6	254·0	279·4

FEET TO MILLIMETRES (*1 foot = 304·800 mm*)

ft	0	1	2	3	4	5	6	7	8	9	10
mm	—	304·8	609·6	914·4	1219·2	1524·0	1828·8	2133·6	2438·4	2743·2	3048·0

POUNDS TO KILOGRAMMES *All values in this table to the nearest gramme*

lb	1	2	3	4	5	6	7	8	9	10	11	12	13	14
kg	0·454	0·907	1·361	1·814	2·268	2·722	3·175	3·629	4·082	4·536	4·990	5·443	5·897	6·350

lb	10	20	30	40	50	60	70	80	90	100
kg	4·536	9·072	13·608	18·144	22·680	27·216	31·751	36·287	40·823	45·359

inches	mm	inches	mm	inches	mm	inches	mm
0·60	15·24	0·70	17·78	0·80	20·32	0·90	22·86
0·61	15·49	0·71	18·03	0·81	20·57	0·91	23·11
0·62	15·75	0·72	18·29	0·82	20·83	0·92	23·37
0·63	16·00	0·73	18·54	0·83	21·08	0·93	23·62
0·64	16·26	0·74	18·80	0·84	21·34	0·94	23·88
0·65	16·51	0·75	19·05	0·85	21·59	0·95	24·13
0·66	16·76	0·76	19·30	0·86	21·84	0·96	24·38
0·67	17·02	0·77	19·56	0·87	22·10	0·97	24·64
0·68	17·27	0·78	19·81	0·88	22·35	0·98	24·89
0·69	17·53	0·79	20·07	0·89	22·61	0·99	25·15
						1·00	25·40

DEGREES TO RADIANS (180° = π rad)

degrees	radian	degrees	radian	degrees	radian
1°	0·017453	10°	0·174533	100°	1·74533
2°	0·034907	20°	0·349066	110°	1·91986
3°	0·052360	30°	0·523599	120°	2·09440
4°	0·069813	40°	0·698132	130°	2·26893
5°	0·087266	50°	0·872665	140°	2·44346
6°	0·104720	60°	1·04720	150°	2·61799
7°	0·122173	70°	1·22173	160°	2·79253
8°	0·139626	80°	1·39626	170°	2·96706
9°	0·157080	90°	1·57080	180°	3·14159

Bibliography

Applied Anthropometry

A.G.A.R.D., *Anthropometry and Human Engineering*, Butterworth Scientific Publications, 1955

Akerblom, B., *Standing and Sitting Posture*, A. B. Nordiske Bokhandelm, Stockholm, 1948

Barkla, D. M., "Chair Angles, Duration of Sitting and Comfort Ratings", *Ergonomics*, No. 7, 1964

Barkla, D. M., "The Estimation of Body Measurements of British Population in Relation to Seating", *Ergonomics*, No. 4

Barter, J. T., A Statistical Evaluation of Joint Range Data, Aero Medical Laboratory USAF Wright Air Development Center, TR-S7-311, 1957

Bayer, L. M. and Gray, H., Anthropometric Standards for Working Women, *Human Biology*, No. 6, 1934

Bedford, T. and Warner, C. G., "Strength Tests: Observations on the Effects of Posture on the Strength of Pull", *Lancet*, No. 2, 1937

Branton, P., *The Comfort of Easy Chairs*, F.I.B.R.A. Report No. 22, 1966

Branton, P., *Seating in Industry*, Ministry of Technology, 1966

British Boot, Shoe and Allied Trade, *Dimensions of Children's Feet*, Research Report No. 98, 1947

British Standards Institution, *Anatomical, Physiological and Anthropometric Principles in the Design of Office Chairs and Tables*, B.S.3044, 1958

British Standards Institution, *Anthropometric Recommendations for Dimensions for Non-Adjustable Office Chairs and Tables*, B.S.3079, 1959

British Standards Institution, *Anthropometric Recommendations for Dimensions of Office Machine Operators*, B.S.3404, 1959

Carlsöö, S., Muscle load in working postures. *Ergonomics* No. 4 1961

Chapanis, A. R. E., *Research Techniques in Human Engineering*, John Hopkins Press, 1959

Chapanis, A. R. E., *Lectures on Man and Machines. An Introduction to Human Engineering*, John Hopkins University Report 166-1-19, 1947

Clapham, J. R. C., and Durdin, J., *Chairs and Sitting*, Proc. Ergonomics Research Soc. No. 2, 1954

Clements, E. M. and Baverstock, M. and Pickett, K., *The Anthropometric Considerations which Underlie Sizes of Childrens Clothing*, University of Birmingham, Department of Anatomy, Replort No. 1, 1955
Damon, A. and McFarland, R. A., "The Physique of Bus and Truck Drivers: With a Review of Occupational Anthropology", *American Journal of Physical Anthropology,* No. 13, 1955
Darcus, H. D. and Weddell, A. G. M., "Some Anatomical and Physiological Principles Concerned in Seats for Naval War Weapons", *British Medical Bulletin*, No. 5, 1947
Dempster, W. T., The Anthropometry of Body Action, *American New York Academy of Science,* No. 63, 1955
Durand, V. and Grandjean, E., "Sitting Habits of Office Employees," *Ergonomics,* No. 6, 1963
Fisher, M. B. and Birren, J. E., "Standardization of a Test of Hand Strength", *Journal of Applied Psychology*, No. 30, 1946
Floyd, W. F. and Roberts, D. F., "Anatomical and Physiological Principles in Chair and Table Design," *Ergonomics*, No. 2, 1958
Floyd, W. F. and Welford, A. T., *Symposium on Human Factors in Equipment Design,* H. K. Lewis and Co. Ltd., 1954
Hertzberg, H. T. E., *Some Contributions of Applied Physical Anthropology to Human Engineering*, American New York Academy of Science, No. 63, 1955
Hertzberg, H. T. E., *Anthropometry of Flying Personnel*, U.S.A.F. Wright Paterson Air Force Development Centre, TR. 53-32, 1954
Hick, W. E., "Why the Human Operator?" *Transactions of the Society of Instrument Technology*, 1952
Hooton, E. A., *A Survey in Seating*, Harvard University, 1955
Hugh-Jones, P., "The Effect of Limb Position in Seated Subjects Their Ability Utilize the Maximum Contractile Force of the Limb Muscles", *Journal of Physiology*, No. 105, 1947
Hunsicker, D. A. and Creey, G., Studies in Human Strength, *Research Quarterly of the American Association for Health, Physical Education and Recreation,* No. 28, 1957
Jones, J. C. and Thornley, D. G. (editors), *Conference on Design Methods*, Pegamon Press, 1963
Karpinos, B. D., "Height and Weight of Selective Service Registrants Processed for Military Service During World War II," *Human Biology*, No. 30, 1958
Keegan, J. J., Evaluation and Improvement of Seats, *Industrial Medicine and Surgery*, No. 31, 1962

Keegan, J. J., "Anthropometry of Chairs", *Architectural Design*, 1954

Keegan, J. J., "Alterations in the Lumbar Curve Related to Posture and Seating", *Journal of Bone and Joint Surgery*, No. 35A, 1953

Krogman, W. M. and Johnson, F. E., Human Mechanics, Aerospace Medical Research Laboratories U.S.A.F. Wright Paterson Air Force Development Centre, AMRL-TDR-63-123, 1963

Martin, W. E., *The Functional Body Measurements of School Age Children*, Chicago National Service Institute Publication, 1954

McConville, J. T. and Hertzberg, H. T. E., *A Study of One Handed Lifting*, Aerospace Medical Research Laboratories U.S.A.F. Wright Paterson Air Force Development Centre, AMRL-TR-66-17, 1966

McCormick, E. J., *Human Engineering*, McGraw-Hill Book Co. Inc., 1957

McFarland, R. A., "Anthropometry in the Design of Drivers Workspace", *American Journal of Physical Anthropology*, No. 16, 1958

Morant, G. M., "Body Size and Work Spaces," *Proc. Ergonomics Research Soc.*, No. 2, 1954

Morgan, C. T., *Human Engineering to Equipment Design*, McGraw-Hill Book Co. Inc., 1963

Murrell, K. F. H., *Ergonomics*, Chapman and Hall, 1965

O'Brien, R. and Shelton, W. C., *Women's Measurements for Garments and Pattern Construction*, Department of Agriculture Washington D.C., Publication No. 454, 1941

O'Donovan, B., "Seating Dimensions Theory and Practice", *Design*, January, 1961

Parnell, R. W., Physical Body Measurements, *Transactions Assistant Industrial Medical Officers*, 1951

Roberts, D. F., *Human and Inhuman Factors in Office Design*, Paper for Royal Society of Health, 1962

Roberts, D. F., "Functional Anthropometry of Elderly Women," *Ergonomics*, No. 3, 1960

Roberts, D. F., "Industrial Applications of Body Measurements," *American Anthropology*, No. 58, 1956

Taylor, C. L. and Boelter, L. M. K., "Biotechnology; A New Fundamental in the Training of Engineers," *Science*, No. 105, 1946

Welford, A. T., "*Ergonomics of Automation*," Department of Scientific and Industrial Research, H.M.S.O., 1960

Whitney, R. J., The Strength of the Lifting Action in Man, *Ergonomics*, No. 1, 1958
Woodson, W. E., *Human Engineering Guide for Equipment Designers*, Cambridge University Press, 1954

Growth, Ageing and Physique

Barton, W. H. and Hunt, E. E., "Somatotype and Adolescence in Boys: A Longtitudinal Study," *Human Biology*, No. 4, 1962
Belbin, R. M., "Older People and Heavy Work", *British Journal of Industrial Medicine*, No. 12, 1955
Birren, J., *Handbook of Ageing and the Individual*, Chicago University of Chicago Press, 1959
Bourne, G. H., Editor, *Structural Aspects of Ageing*, Pitman, 1961
Brozek, J. Hunt, E. E. and Skerjl, B., "Subcutaneous Fat and Age Changes in Body Build and Body Form in Women", *American Journal of Physical Anthropology*, No. 11, 1953
Edwards, D. A., "Differences in the Distribution of Subcutaneous Fat with Sex and Maturity", *Clinical Science*, No. 10, 1951
Edwards, D. A., "Observations on the Distribution of Subcutaneous Fat", *Clinical Science*, No. 9, 1959
Gallagher, J. R. and Seltzer, C. C., "Somatotypes of an Adolescent Group", *American Journal of Physical Anthropology*, No. 2, 1946
Jones, H. E., "The Relationship of Strength to Physique", *American Journal of Physical Anthropology*, No. 5, 1947
Kemsley, W. F. F., "Bodyweight at Different Ages and Weights", *Annals Eugenics*, 1952
Tanner, J. M., *The Physique of the Olympic Athlete*, G. Allen and Unwin Ltd., 1964
Tanner, J. M., *Growth at Adolescence*, II edition, Blackwell Scientific Publications 1955
Welford, A. T., *Ageing and Human Skill*, Oxford University Press for Nuffield Foundation, 1958

Human Biology

Harrison, G. A., Editor and others, *Human Biology*, Oxford University Press, 1964
Weiner, J. S. and Lourie, J. A., *Human Biology. A Guide to Field Methods. Handbook* No. 9, Blackwell Scientific Publications, 1969

Anthropometric measuring instrument suppliers

Siber Hegner & Co. Ltd, Talstrasse 14, 8022 Zürich, Switzerland

Siber Hegner & Co. Ltd, 8 West 30th Street, New York 1, NY.

Glossary

Acceleration	The rate of increase of a measurement per unit of time.
Accretion	The process of growth by external addition to cause increase.
Adaptation	The ability to make changes in the physiological system in reaction to noted environmental conditions without detracting from mental or physical abilities.
Allometry	Differential growth of one part of the body to the remainder. It is because of differential growth rates that the body exhibits changes of form and proportion.
Articulation	The relationship of two bodies by means of a moveable joint.
Caudal	Nearer the feet.
Cybernetics	The study of the processes of control, communication, behaviour and learning, induced or created, in natural or engineered systems; and the inter-relationship of the processes of a system with a containing environment.
Dimorphism	The occurrence and display of two forms distinct in proportion and structure amongst animals of the same species.
Dorsal	To the rear of the body.
Dynamic	Relating to a force in action producing a motion.
Ecology	The study of the home conditions of an organism.
Endocrine	*See* Glands.
Environment	The envelope of conditions in which a person lives or works; a geographical or industrial location reckoned in terms of temperature, humidity, air pressure or light and all other external stimuli.

Ergonomics	The study of human mental and physical performance in any work situation in terms of efficiency and competence. Additionally the design of ideal working environments and the allocation of tasks in a man and machine situation.
Epiphysis	A separate centre of ossification from the main portion of a bone, from which it is isolated by cartilage. Present in the growing child, the cartilage disappearing at physical maturity.
Factor	A quantity or circumstance that produces a result.
Fossa	Meaning a ditch. In anatomy applied to a variety of depressions.
Fulcrum	The pivot point of a lever which is favourable for applying force.
Gene	The units of inheritance which help to form the characteristics of an individual.
Genetics	The study of heredity.
Genotype	The constitutional inheritance which reacts with the environment to mould the phenotype.
Gerontology	The branch of medical science that deals with the treatment of the aged.
Glands	Organs which contain secretory cells, of which there are numerous types in the body. The endocrine (or ductless glands) are of no standard type, but are spread throughout the body and play a vital part in the function that promote growth and development.
Habitat	The locality in which an organism can live and leave descendants.
Homeostasis	The ability to continue a steady or optimum condition of life by processes of internal self regulation or adjustment. Self regulation of the internal environment.
Homo	A genus of primates, of the order mammals.

Homo sapien	The human animal, the only living representative of the genus Homo.
Kinesiology	The mechanics of the motion of the human body.
Linea aspera	A prominent ridge on the back of the femur bone for the attachment of muscles. Its prominence is an indication of the power of the leg muscles.
Locomotion	The power of an animal to exert force progressively on the limbs and produce motion.
Mammal	A class of vertebrates. Distinguished by the female suckling young, a unified lower jaw and the possession of body hair at some period.
Metabolism	The bio-chemical and bio-physical processes by which life is maintained in the body, and by which energy is released.
Optimum	The amount of stimulus that allows all the processes of a function tc behave normally.
Ossification	The production of bone; either from cartilage or embryonic tissue.
Paidology (Paedology)	The study of growing children.
Parameter	The construction of a convenient variable in which other variables can be expressed.
Pathology	The branch of medical science that deals with the causes of disease and the changes they produce in the body.
Phenotype	The form that is appearing in childhood and which is due to both genetic and environmental influences.
Plantigrade	The act of using the whole of the sole of the foot in walking.
Prehensile	The capability of grasping or gripping action.
Prone	Having the front portion of the body downwards.
Proportion	The relationship between two quantities or

	sizes. The relationship between the part and the whole.
Puberty	The change of form and function that takes place as a child enters maturity.
Sexual dimorphism	The characteristic differences between males and females of the same species.
Static	Bodies at rest, or the equilibrium of forces when no motion is produced.
Stimulus	That which can effect or excite the activity or function of an organ to produce a material reaction.
Supine	Having the front portion of the body upwards.
Syndesmology	The description and study of articulations in anatomy.
Tactile	The sense of touch. Perception and belief by touch.